COMPUTER
APPLICATIONS
IN
HYDRAULIC
ENGINEERING

HAESTAD
PRESS

Computer Applications in Hydraulic Engineering

This book is published by Haestad Methods, Inc. and is intended for
academic purposes and educational needs.

Haestad Methods, Inc. Trademarks

The following are registered trademarks of Haestad Methods, Inc:
CADmagic, CulvertMaster, CYBERNET, FlowMaster, StormCAD, WaterCAD.

The following are trademarks of Haestad Methods, Inc:
HECPACK, POND-2, POND PACK, SewerCAD, Visual HEC-1, Visual HEC-Pack

Haestad Methods is a registered tradename of Haestad Methods, Inc.

All other brands, company or product names or trademarks belong to their respective holders.

Library of Congress Catalog Card Number: 97-72243
ISBN Number 0-9657580-1-X

Contributing Editors

Haestad Methods Engineering Staff

Michael E. Meadows, Ph.D., P.E.
University of South Carolina

Thomas M. Walski, Ph.D., P.E.
Wilkes University

Contents

Foreword

Since 1979, Haestad Methods, Inc. has been developing hydrology and hydraulic software as a design tool for civil engineers, providing programs and technical support to tens of thousands of professional civil engineers, modelers, and universities.

Along the way, we have learned a great deal about our clients, their professional backgrounds, and also their educational backgrounds. We also offer continuing education courses for professionals who need to get up-to-speed quickly with various numerical methods and practices.

Why is this important?

Our experience has shown us a great deal about the areas where engineers are being trained, and it has also revealed a significant gap in this training - the link between hydraulic theory and practical computer applications.

There are hundreds of textbooks that offer enormous detail in the areas of engineering history, equation derivations, and hand calculation methods. There are also hundreds of published theses and articles that deal with computer applications - unfortunately, most of them are highly research-oriented, and are usually tied to a specific case study or an unusual set of circumstances. Both of these publication types are very important to the civil engineering industry, and it is still very important to understand the concepts of hydraulic modeling (and learn the problem solving skills and procedures that are associated with performing calculations longhand).

This publication is intended as an introduction to more practical applications of engineering software, demonstrating the types of situations that an engineer will most likely come across on a daily basis in the real world. It shows the true benefits of computer software - increased efficiency, better flexibility, and - most importantly - an increased ability to try different and better designs.

It is our hope that engineers, technicians, and students will find this book to be challenging, but also easy to understand and very practical. Combined with standard hydraulic references, we believe that this provides the tools needed to successfully proceed with a career in the field of hydrology and hydraulics.

Chapter 1
Basic Hydraulic Principles

1.1 General Flow Characteristics

In hydraulics, as with any technical topic, full understanding cannot come without first becoming familiar with the basic terminology and governing principles. Several of these basic concepts, discussed in the following pages, lay the foundation for the more complex analyses that are presented in later chapters.

Flow Conveyance

As we should all know by now, water travels downhill (unless forced to do otherwise) until it reaches a leveling point, such as an ocean. This tendency is facilitated by the presence of natural conveyance channels such as brooks, streams, and rivers. The water's voyage may also be aided by man-made structures such as drainage swales, pipes, culverts, and canals. Although engineering design is most commonly interested in the characteristics of the man-made features, the hydraulic concepts can be applied equally well to natural materials.

Area, Wetted Perimeter, and Hydraulic Radius

Quite simply, the term *area* refers to the cross-sectional area of flow within the channel. When a channel has a consistent cross-sectional shape, slope, and roughness, it is called a *prismatic* channel.

If the channel is open to the atmosphere, such as a culvert flowing only partially full or a natural river, it is said to be *open channel flow* or *free surface flow*. If the channel is flowing completely full, such as a water pipe, it is said to be *full flow*. *Pressure flow* is a special type of full flow, where forces on the fluid cause it to push against the top of the channel as well as the bottom and sides. For example, these forces may be from the weight of a column of water in a backed-up sewer manhole, or from an elevated storage tank.

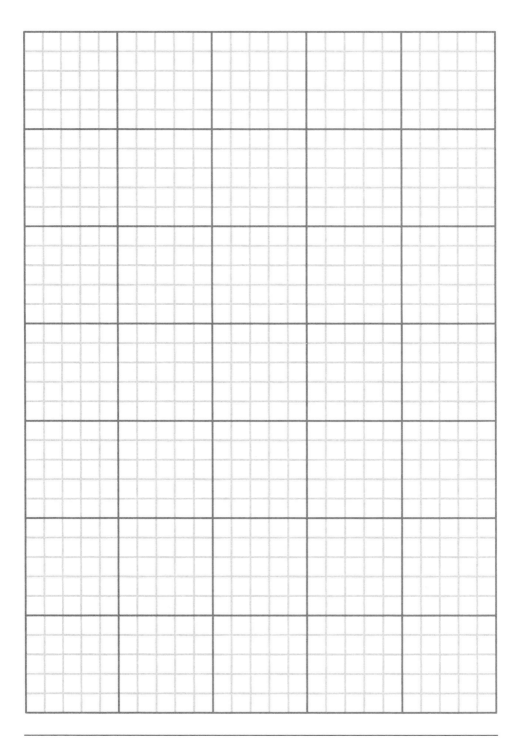

A section's **wetted perimeter** is defined as the portion of the flow area's perimeter that is in contact with the conveyance channel (as opposed to the portion of the perimeter that is open to the atmosphere). This is shown in the following figure.

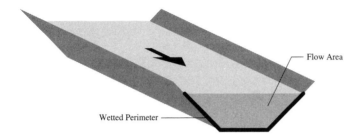

Figure 1-1: Flow Area and Wetted Perimeter

The **hydraulic radius** of a section is not a directly measurable characteristic, but is frequently used during calculations. It is defined as the area divided by the wetted perimeter, and therefore has units of length.

For specific flow conditions, the hydraulic radius can often be related directly to the geometric properties of the channel. For example, the hydraulic radius of a full circular pipe (such as a pressure pipe) can be directly computed as:

$$R = \frac{A}{P_w}, \quad R_{circular} = \frac{\pi \cdot D^2/4}{\pi \cdot D} = \frac{D}{4}$$

where   R   is the hydraulic radius (ft or m)
            A   is the cross-sectional area (ft^{2}or m^2)
            P_w is the wetted perimeter (ft or m)
            D   is the pipe diameter (ft or m)

Velocity

The **velocity** of a section is not constant throughout the area - instead, it varies with location. Where the fluid is in contact with the conduit wall, the velocity is zero.

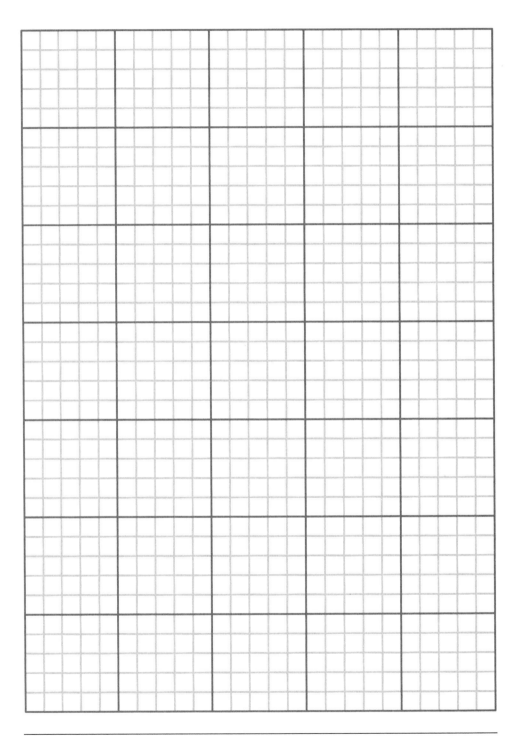

Haestad Methods, Inc.

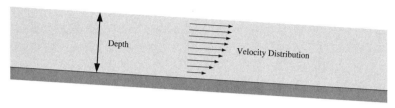

Longitudinal Section (Profile)

Figure 1-2: Velocity Distribution

This adds a great deal of complication to any sort of hydraulic analysis, so the engineer is usually concerned with the mean (average) velocity of the section for analysis purposes. This average velocity is defined as the total flowrate divided by the cross-sectional area, and is in units of length per time.

$$V = Q / A$$

where   V   is the mean velocity (ft/s or m/s)
        Q   is the flowrate (ft^3/s or m^3/s)
        A   is the area (ft^2 or m^2)

Steady Flow

When speaking in terms of flow, the word "steady" indicates that a constant flowrate is assumed throughout the analysis. For most hydraulic calculations, this assumption is completely reasonable - a minimal increase in model accuracy does not warrant the time and effort that would be required to perform an analysis with changing (unsteady) flows.

Even for tributary and river networks, stormwater sewers, and other collection systems where the flowrates do change throughout the system, the network can often be broken into segments for analysis that can be analyzed separately under steady flow conditions.

Laminar Flow, Turbulent Flow, and Reynolds Number

Laminar flow is a flow type that is characterized by smooth flow lines, such as the appearance of maple syrup being poured. *Turbulent* flow, on the other hand, does not have smooth flow lines at all. Turbulent flow is characterized by the formation of eddies within the flow, resulting in continuous mixing throughout the section.

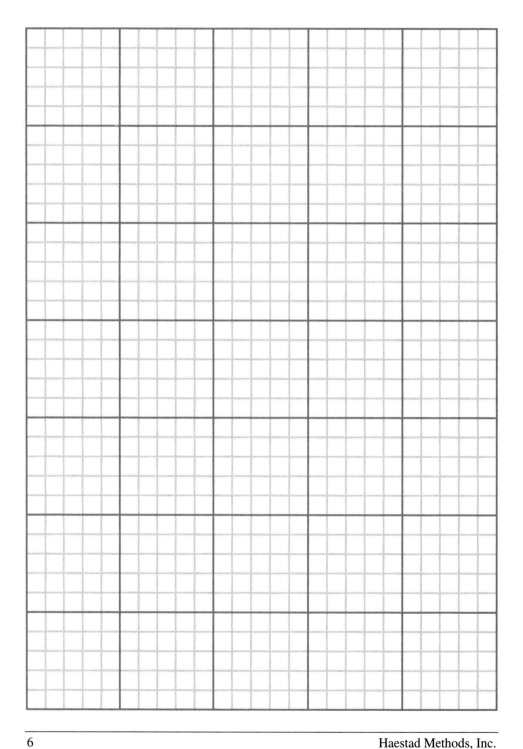

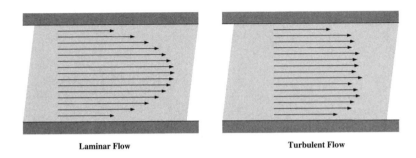

Laminar Flow   Turbulent Flow

Figure 1-3: Instantaneous Velocity Distributions for Laminar and Turbulent Flow

Note that the eddies result in varying velocity directions as well as magnitudes (not depicted above for simplicity). Note also that the eddies at times contribute to the velocity of a given particle in the direction of flow, and at other times detract from it. The result is that velocity distributions captured at different times are likely to be quite different from each other, and both will be far rougher than the velocity distribution of a laminar flow section.

By strict interpretation, the changing velocities in turbulent flow would be considered unsteady. Over time, however, the average velocity at any given point within the section is essentially constant, so the flow is assumed to be steady.

Interestingly, the velocity at any given point within the turbulent section will be closer to the mean velocity of the entire section than for laminar flow conditions. This occurs because of the continuous mixing of flow - particularly the mixing of low-velocity flow near the conduit walls with the higher-velocity flow towards the center.

To classify flow as either turbulent or laminar, there is an index called the ***Reynolds number***. It is computed as:

$$R_e = \frac{4VR}{\nu}$$

where R_e is the Reynolds Number
        V   is the mean velocity (ft/s or m/s)
        R   is the hydraulic radius (ft or m)
        ν is the kinematic viscosity, available from most fluid tables (ft^2/s or m^2/s)

If the Reynolds number is below 2,000, the flow is generally laminar. If the Reynolds number is above 4,000, the flow is generally turbulent. Between 2,000 and 4,000, the flow may be either laminar or turbulent depending on other factors.

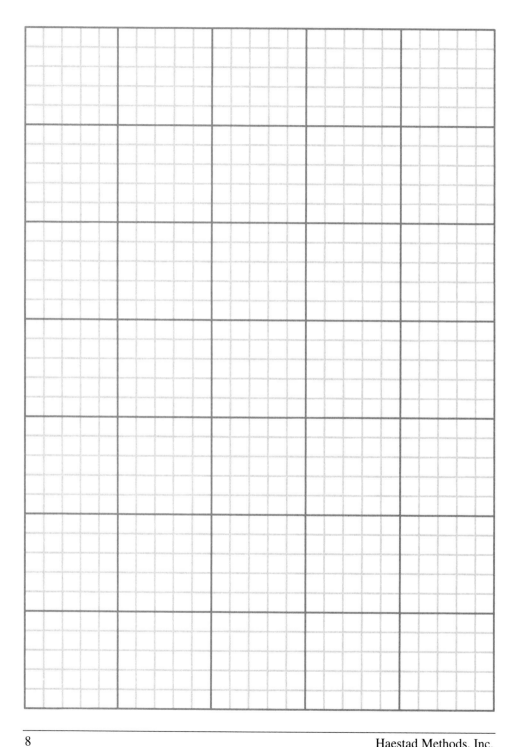

Haestad Methods, Inc.

Example 1-1: Flow Characteristics

A rectangular concrete channel is 3 meters wide and 2 meters high. The water in the channel is 1.5 meters deep, and is flowing at a rate of 30 m³/s. Determine the flow area, wetted perimeter, and hydraulic radius. Is the flow laminar or turbulent?

Solution: From the section's shape (rectangular), we can easily calculate the area as the rectangle's width multiplied by its depth. Note that the depth used should be the actual depth of flow, not the total height of the cross-section. The wetted perimeter can also be found easily through simple geometry.

$$A = 3.0 \text{ m} \bullet 1.5 \text{ m} = 4.5 \text{ m}^2$$

$$P_w = 3.0 \text{ m} + 2\bullet(1.5 \text{ m}) = 6.0 \text{ m}$$

$$R = A / P_w = 4.5 \text{ m}^2 / 6.0 \text{ m} = 0.75 \text{ m}$$

In order to determine whether or not the flow is turbulent, we must determine the Reynolds number. To do this, we must first find the velocity of the section, and a value for the kinematic viscosity.

$$V = Q / A = 30 \text{ m}^3/\text{s} / 4.5 \text{ m}^2 = 6.67 \text{ m/s}$$

From fluids tables, we find that the kinematic viscosity for water at 20°C is $1.00\text{x}10^{-6}$ m²/s. Substituting these values into the formula to compute Reynolds number results in

$$R_e = (4 \bullet 6.67 \text{ m/s} \bullet 0.75 \text{ m}) / (1.00\text{x}10^{-6}) = 20,000,000$$

This is well above the Reynolds number minimum of 4,000 for turbulent flow.

1.2 Energy

The Energy Principle

The first law of thermodynamics states that for any given system, the change in energy (ΔE) is equal to the difference between the heat transferred to the system (Q) and the work done by the system on its surroundings (W) during a given time interval.

The energy referred to in this principle represents the total energy of the system – the sum of the potential energy, kinetic energy, and internal (molecular) forms of energy such as electrical and chemical energy. While these internal energies may be significant for thermodynamic analysis, they are commonly neglected in hydraulic analysis because of their relatively small magnitude.

In hydraulic applications, energy values are often converted into units of energy per unit weight, resulting in units of length. Using these length equivalents gives engineers a

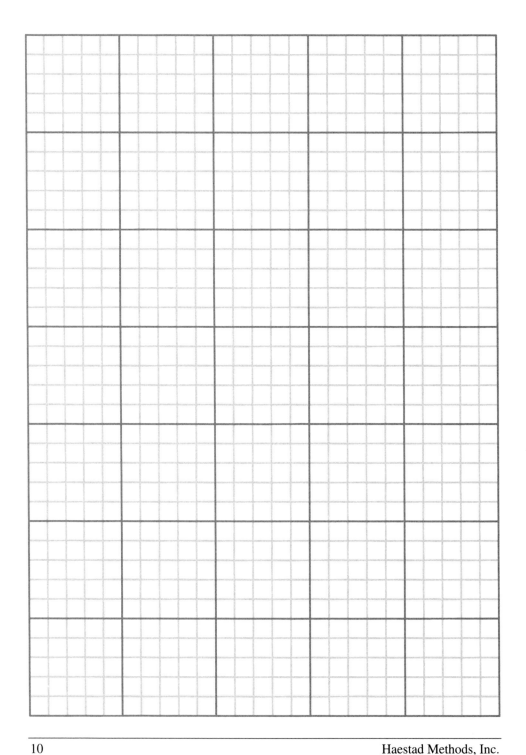

Haestad Methods, Inc.

better "feel" for the resulting behavior of the system. When using these length equivalents, the engineer is expressing the energy of the system in terms of "head". The energy at any point within a hydraulic system is often represented in three parts:

- Pressure Head p/γ

- Elevation Head z

- Velocity Head $V^2/2g$

where p is the pressure (lbs/ft^2 or N/m^2)
        γ is the specific weight (lbs/ft^3 or N/m^3)
        z is the elevation (ft or m)
        V is the velocity (ft/s or m/s)

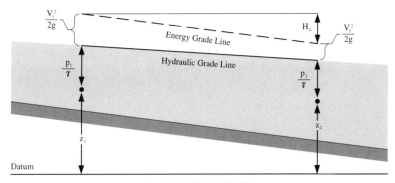

Longitudinal Section (Profile)

Figure 1-4: The Energy Principle

Note that a point on the surface of an open channel will have a pressure head of zero (but a higher elevation head than a point selected at the same station but at the bottom of the channel).

The Energy Equation

In addition to pressure head, elevation head, and velocity head, there may also be energy added to a system (such as by a pump) and energy removed from the system (due to friction or other disturbances). These changes in energy are referred to as head gains and head losses, respectively. Balancing the energy across any two points in the system, we then obtain the energy equation:

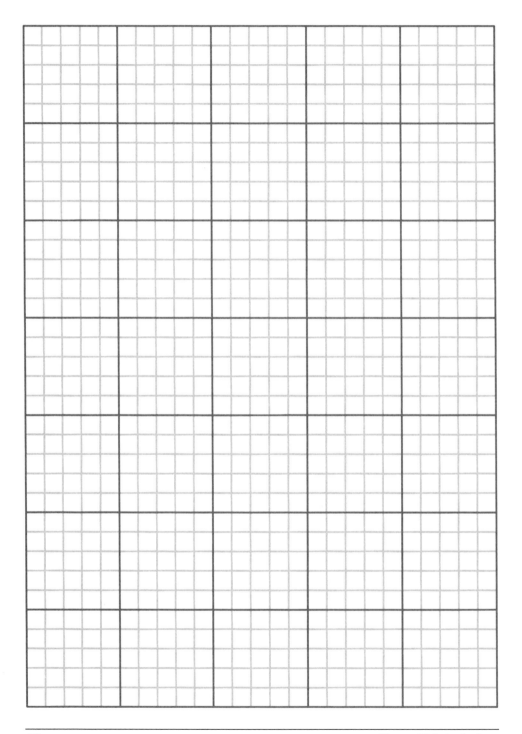

Haestad Methods, Inc.

$$\frac{p_1}{\gamma} + z_1 + \frac{V_1^2}{2g} + H_G = \frac{p_2}{\gamma} + z_2 + \frac{V_2^2}{2g} + H_L$$

where p is the pressure (lb/ft^2 or N/m^2)
        γ is the specific weight of the fluid (lb/ft^3 or N/m^3)
        z is the elevation at the centroid (ft or m)
        V is the fluid velocity (ft/s or m/s)
        g is gravitational acceleration (ft/s^2 or m/s^2)
        H_G is the head gain, such as from a pump (ft or m)
        H_L is the combined head loss (ft or m)

Hydraulic Grade

The **hydraulic grade** is the sum of the pressure head (p/γ) and elevation head (z). For open channel flow, the hydraulic grade is the water surface elevation (since the pressure head is zero). For a pressure pipe, the hydraulic grade represents the height to which a water column would rise in a piezometer (a tube rising from the pipe). When plotted in profile versus the length of the conveyance section, this is often referred to as the hydraulic grade line, or HGL.

Energy Grade

The **energy grade** is the sum of the hydraulic grade and the velocity head ($V^2/2g$). This is the height to which a column of water would rise in a Pitot tube (an apparatus similar to a piezometer, but also accounting for fluid velocity). When plotted in profile, this is often referred to as the energy grade line, or EGL. At a lake or reservoir, where the velocity is essentially zero, the EGL is equal to the HGL.

Energy Losses and Gains

Energy losses (H_L) in a system may be due to a combination of several factors. The primary cause of energy loss is usually due to internal friction between fluid particles traveling at different velocities. Secondary causes of energy loss are due to localized areas of increased turbulence and disruption of the streamlines, such as disruptions from valves and other fittings in a pressure pipe, or disruptions such as a changing section shape in a river.

The rate at which energy is lost along a given length of channel is called the friction slope, and is usually presented as a unitless value, or in units of length per length (ft/ft, m/m, etc.).

The only way for a substantial amount of energy to be added to a system is through a manmade device such as a pump. Pumps are discussed in more detail in Chapter 4.

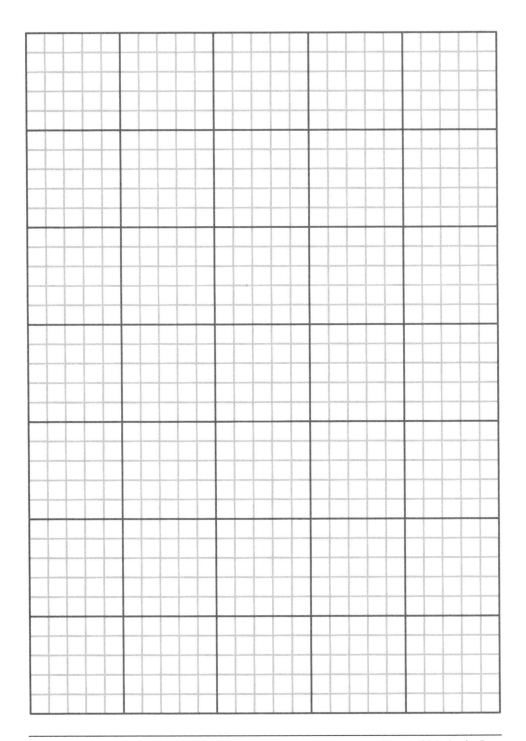

Haestad Methods, Inc.

Example 1-2: Energy Principles

A 48-inch diameter transmission pipe carries 2,000 gallons per minute from an elevated storage tank with a water surface elevation of 1,764 feet. Two miles from the tank, at an elevation of 1,423 feet, a pressure meter reads 85 psi. If there are no pumps between the tank and the meter location, what is the rate of headloss in the pipe?

Solution: We can begin by simplifying the energy equation. We can assume that the velocity within the tank is negligible, and we can also discount the pressure head at the tank since it is open to the atmosphere. Rewriting the energy equation and entering the known values, we can easily solve for the amount of headloss. The only value that we still need to calculate is the velocity, which can be found from the flowrate and pipe diameter.

$$Q = 2,000 \text{ gal/min} \bullet (1 \text{ ft}^3/\text{s} / 448.86 \text{ gal/min}) = 4.46 \text{ ft}^3/\text{s}$$

$$A = \pi \bullet (2 \text{ ft})^2 = 12.57 \text{ ft}^2$$

$$V = Q / A = 4.46 \text{ ft}^3/\text{s} / 12.57 \text{ ft}^2 = 0.35 \text{ ft/s}$$

$$V^2 / 2g = (0.35 \text{ ft/s})^2 / (2 \bullet 32.2 \text{ ft/s}^2) = 0.002 \text{ ft (negligible)}$$

This simplifies the energy equation even further, and we can now solve for headloss as

$$H_L = 1764 \text{ ft} - 1423 \text{ ft} - (85 \text{ lb/in}^2)(144 \text{ in}^2/\text{ft}^2)/62.4 \text{ lb/ft}^3 = 145 \text{ ft}$$

We still aren't quite done yet, though, since the original problem statement asked for a rate of headloss. This is usually presented as a friction slope, or in feet of loss per 1000 feet of pipe.

$$\text{Friction slope} = 145 \text{ ft} / (2 \bullet 5,280 \text{ ft}) = 0.014 \text{ ft/ft, or 14 feet per thousand.}$$

1.3 Friction Losses

There are many equations that approximate friction losses associated with the flow of a liquid through a given section. Commonly used methods include:

- Manning's equation

- Chezy's (Kutter's) equation

- Hazen-Williams equation

- Darcy-Weisbach (Colebrook-White) equation

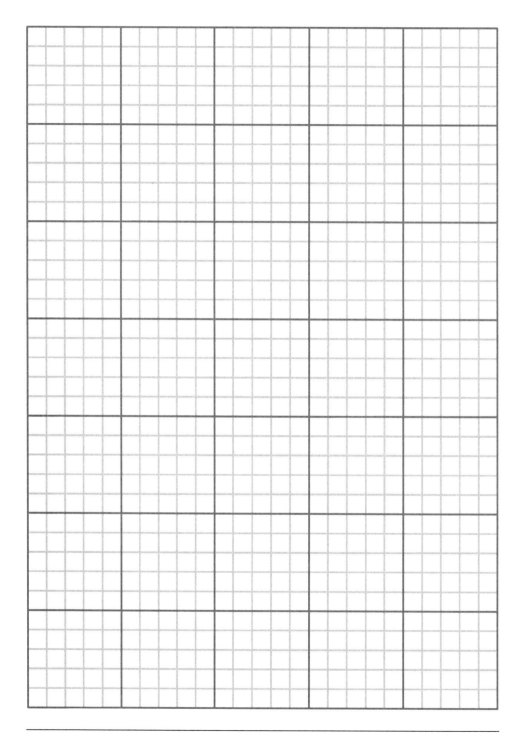

Haestad Methods, Inc.

These equations can all be described by a generalized friction equation:

$$V = kCR^x S^y$$

where   V   is the mean velocity
           C   is a flow resistance factor
           R   is the hydraulic radius
           S   is the friction slope
           x, y are exponents
           k   is a factor to account for empirical constants, unit conversion, etc.

The lining material of the flow channel usually determines the flow resistance (or roughness) factor, C. However, the ultimate value of the C component may be a function of the channel shape, depth, and fluid velocity.

Manning's Equation

Manning's equation is the most commonly used open channel flow equation. The roughness component, C, is typically assumed to be constant over the full range of flows and is represented by a Manning's roughness value, n. These n values have been experimentally determined for various types of material, and should not be used with fluids other than water. Manning's equation is as follows:

$$V = \frac{k}{n} R^{2/3} S^{1/2}$$

where   V   is the mean velocity (ft/s or m/s)
           k   is 1.49 for U.S. Standard units, or 1.00 for S.I. units
           n   is the Manning's roughness value
           R   is the hydraulic radius (ft or m)
           S   is the friction slope

Chezy's (Kutter's) Equation

The Chezy equation, in conjunction with Kutter's equation, is widely used in sanitary sewer design and analysis. The roughness component, C, is a function of the hydraulic radius, friction slope, and lining material of the channel. The Chezy equation is as follows:

$$V = C\sqrt{RS}$$

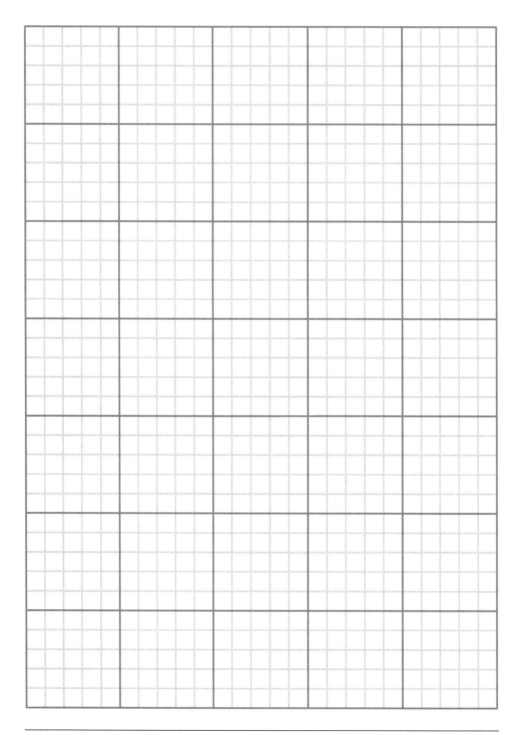

where V is the mean velocity (ft/s or m/s)
        C is the roughness coefficient (see following calculation)
        R is the hydraulic radius (ft or m)
        S is the friction slope

The roughness coefficient, C is related to Kutter's n through the Kutter's equation. Note that the values of n in Kutter's equation are actually Manning's n coefficients.

U.S. Standard Units

$$C = \frac{41.65 + \dfrac{0.00281}{S} + \dfrac{1.811}{n}}{1 + \dfrac{\left(41.65 + \dfrac{0.00281}{S}\right)n}{\sqrt{R}}}$$

S.I. Units:

$$C = \frac{23 + \dfrac{0.00155}{S} + \dfrac{1}{n}}{1 + \dfrac{\left(23 + \dfrac{0.00155}{S}\right)n}{\sqrt{R}}}$$

where C is the roughness coefficient
        n is the Manning's roughness value
        R is the hydraulic radius (ft or m)
        S is the friction slope

Hazen-Williams Equation

The Hazen-Williams equation is most frequently used in the design and analysis of pressure pipe systems. The equation was developed experimentally, and therefore should not be used for fluids other than water (and only within temperatures normally experienced in potable water systems). The Hazen-Williams equation is as follows:

$$V = kCR^{0.63}S^{0.54}$$

where V is the mean velocity (ft/s or m/s)
        k is 1.32 for U.S. Standard units, or 0.85 for S.I. units
        C is the Hazen-Williams roughness coefficient
        R is the hydraulic radius (ft or m)
        S is the friction slope

Darcy-Weisbach (Colebrook-White) Equation

The Darcy-Weisbach Equation is a theoretically based equation commonly used in the analysis of pressure pipe systems. It applies equally well to any flow rate and any incompressible fluid, and is general enough that it can be applied to open channel flow systems. In fact, the ASCE Task Force on Friction Factors in Open Channels (1963) supported the use of the Darcy-Weisbach equation for free-surface flows. This recommendation has not yet been widely accepted since the solution to the equation is

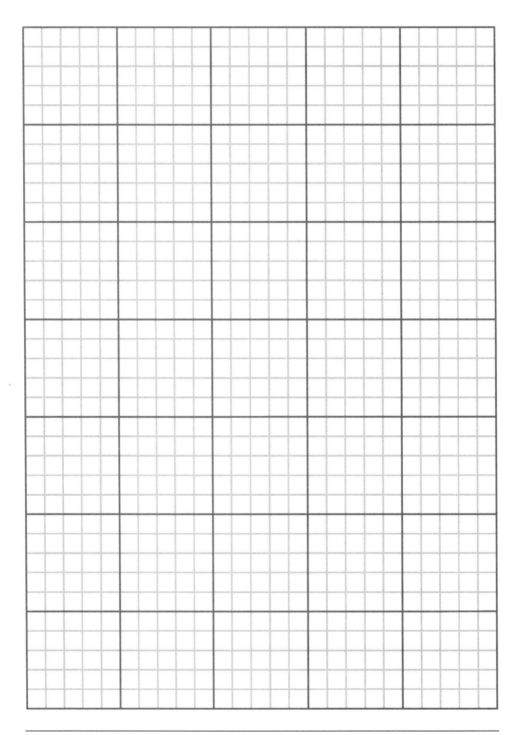

Haestad Methods, Inc.

difficult and not easily computed using simple desktop methods. With the increasing availability of computer solutions, the Darcy-Weisbach equation will likely gain greater acceptance since it successfully models the variability of effective channel roughness with channel material, geometry, and velocity.

The roughness component in the Darcy-Weisbach equation is a function of both the channel material and the Reynolds Number, which varies with velocity and hydraulic radius.

$$V = \sqrt{\frac{8g}{f} RS}$$

where   V   is the flow velocity (ft/s or m/s)
            g   is gravitational acceleration (ft/s^2 or m/s^2)
            f is the Darcy-Weisbach friction factor
            R   is the hydraulic radius (ft or m)
            S   is the friction slope

The Darcy-Weisbach friction factor, f, can be found using the Colebrook equation for fully developed turbulent flow, as follows:

Free Surface   **Full Flow (Closed Conduit)**

$$\frac{1}{\sqrt{f}} = -2\log\left(\frac{k}{12R} + \frac{2.51}{R_e\sqrt{f}}\right) \qquad \frac{1}{\sqrt{f}} = -2\log\left(\frac{k}{14.8R} + \frac{2.51}{R_e\sqrt{f}}\right)$$

where   k   is the roughness height (ft or m)
            R   is the hydraulic radius (ft or m)
            R_e is the Reynolds Number

This iterative search for the correct value of f can become quite time-consuming for hand computations, and even for computerized solutions of many pipes. Another method, developed by Swamme and Jain, solves directly for f in full flowing circular pipes. This equation is as follows:

$$f = \frac{1.325}{\left[\log_e\left(\frac{k}{3.7D} + \frac{5.74}{R_e^{0.9}}\right)\right]}$$

where f is the friction factor
            k   is the roughness height (ft or m)
            D   is the pipe diameter (ft or m)
            R_e is the Reynolds Number

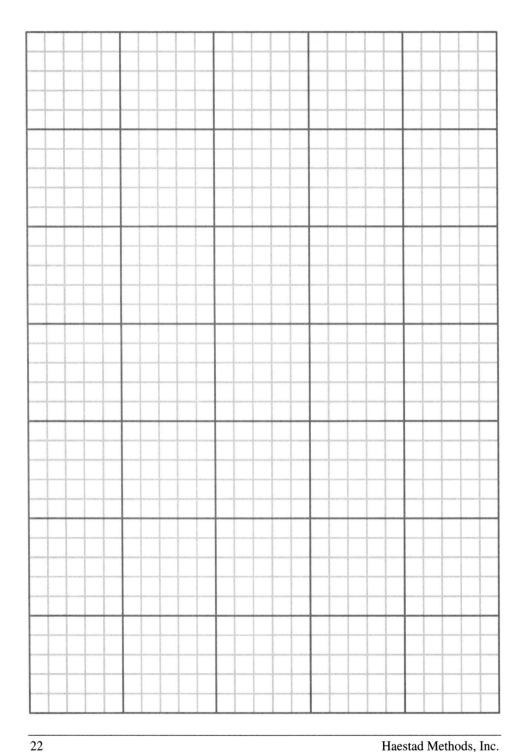

Typical Roughness Factors

Typical pipe roughness values for all of these methods are shown in the following table. These values will vary depending on the manufacturer, workmanship, age, and other factors. For this reason, the following table should be used as a guideline only.

Material	Manning's Coefficient n	Hazen-Williams C	Darcy-Weisbach Roughness Height k (mm)	Darcy-Weisbach Roughness Height k (ft)
Asbestos cement	0.011	140	0.0015	0.000005
Brass	0.011	135	0.0015	0.000005
Brick	0.015	100	0.6	0.002
Cast-iron, new	0.012	130	0.26	0.00085
Concrete:				
Steel forms	0.011	140	0.18	0.006
Wooden forms	0.015	120	0.6	0.002
Centrifugally spun	0.013	135	0.36	0.0012
Copper	0.011	135	0.0015	0.000005
Corrugated metal	0.022	-----	45	0.15
Galvanized iron	0.016	120	0.15	0.0005
Glass	0.011	140	0.0015	0.000005
Lead	0.011	135	0.0015	0.000005
Plastic	0.009	150	0.0015	0.000005
Steel:				
Coal-tar enamel	0.010	148	0.0048	0.000016
New unlined	0.011	145	0.045	0.00015
Riveted	0.019	110	0.9	0.003
Wood stave	0.012	120	0.18	0.0006

Table 1-1: Typical Roughness Coefficients

1.4 Pressure Flow

For pipes flowing full, many of the friction loss calculations are greatly simplified, since the flow area, wetted perimeter, and hydraulic radius are all known as functions of pipe radius (or diameter).

There is much more information presented about pressure piping systems in Chapter 4, including further discussion on pumping systems, minor losses, and network analysis.

Example 1-3: Pressure Pipe Friction Losses

Use the FlowMaster program to compare headloss computed by the Hazen-Williams equation to the headloss computed by the Darcy-Weisbach equation for a pressure pipe

---

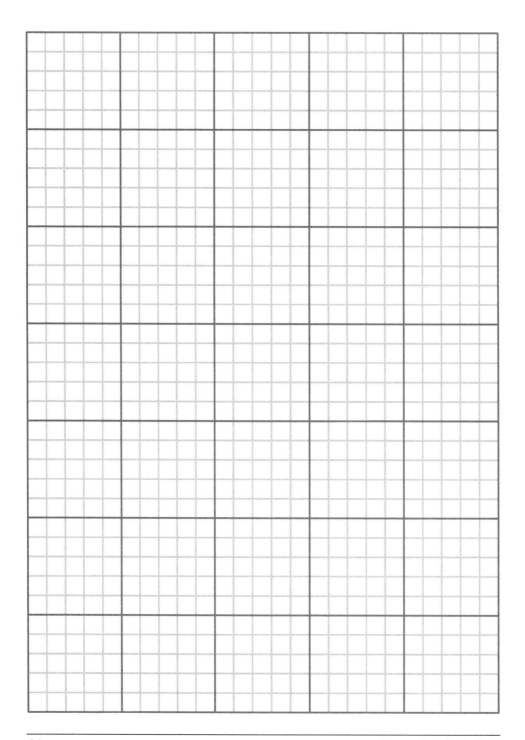

Haestad Methods, Inc.

having the following characteristics: cast iron pipe (new) one mile in length, 12" in diameter at a flowrate of 1200 gallons per minute (with water at 65°F).

Solution: If you think that there is not enough information provided, re-read the problem statement. Although there are no elevations or pressures given, these values are not needed in order to determine the headloss in the pipe. Setting up FlowMaster to solve for the "Elevation at 1" allows us to use zero elevation and zero pressure assumptions, and fill in the rest of the pipe characteristics.

For the Hazen-Williams equation, a C coefficient of 130 is assumed. This results in 18.8 feet of headloss (which agrees with the computed 18.8 elevation at point 1). Using Darcy-Weisbach, a roughness height of 0.00085 feet is assumed. The solution indicates a headloss of 18.9 feet, only one tenth of a foot difference from the value predicted by Hazen-Williams.

Discussion: If the same system is analyzed with 2,000 to 3,000 gallons per minute of flow, however, the headloss difference becomes almost ten feet!

Why such a big difference? For starters, the two methodologies are completely unrelated, and the estimated roughness coefficients were taken from a list of approximate values. If the Hazen-Williams equation is used with a roughness value of 125, the results are much closer. This should emphasize the fact that models are only as good as the data that is being fed to them, and the engineer needs to fully understand all of the assumptions that are being made before accepting the results.

1.5 Open Channel Flow

Open channel flow is more complicated than pressure flow, because the flow area, wetted perimeter, and hydraulic radius may not be constant like they are in full-flow piping. Because of this considerable difference between the flow types, there are some additional characteristics that become important when dealing with open channel flow.

Uniform Flow

Uniform flow refers to the hydraulic condition where the discharge and cross-sectional area (and therefore velocity) are constant throughout the length of the channel. For a pipe flowing full, all that this requires is that the pipe be straight and have no contractions or expansions. For an open channel, there are a few additional interesting points:

- The depth of flow must be constant, so the hydraulic grade line must be parallel to the actual channel slope. This depth of flow is called ***normal*** depth.

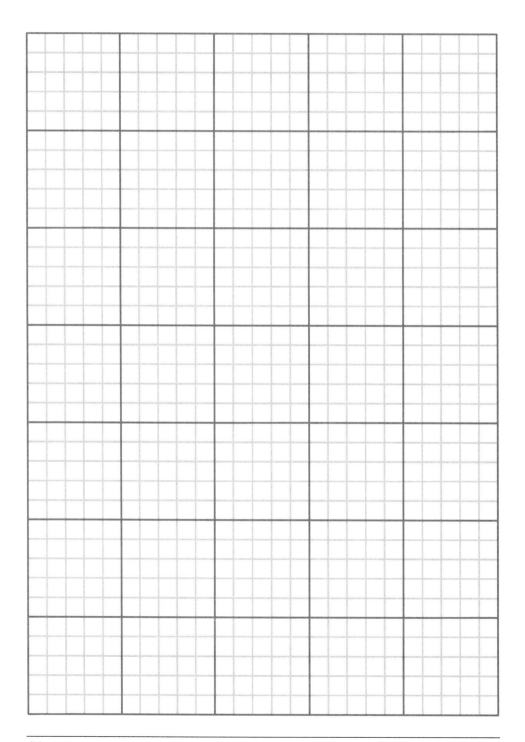

Haestad Methods, Inc.

- Since there is a constant velocity, the velocity head does not change through the length of the section - and therefore the energy grade line is parallel to both the hydraulic grade line and the channel slope.

In channels that are prismatic, the flow conditions will typically approach uniform flow if the channel is sufficiently long. When this occurs, the net force on the fluid approaches zero - the gravitational force is equal to the opposing friction forces from the channel bottom and walls.

Example 1-4: Uniform Flow

A concrete trapezoidal channel has a bottom width of 4 meters and 45° sideslopes. If the channel is on a 1% slope and is flowing 1 meter deep throughout its length, how much flow is being carried (use Manning's equation)? How much flow would the same channel carry if it were squared off at 4 meters instead of trapezoidal?

Solution: Since the channel is flowing at the same depth throughout, we can assume that normal depth has been achieved (so the friction slope is equal to the channel slope). We will assume a Manning's n of 0.013 for concrete.

From the trapezoidal geometry, we can easily calculate the area and wetted perimeter, and then the hydraulic radius, as follows:

$$A = (4 \text{ m} \bullet 1 \text{ m}) + 2 \bullet (0.5 \bullet 1 \text{ m} \bullet 1 \text{ m}) = 5.00 \text{ m}^2$$

$$P_w = 4 \text{ m} + 2 \bullet (1 \text{ m} \bullet 2^{0.5}) = 6.83 \text{ m}$$

$$R = A / P_w = 5.00 \text{ m}^2 / 6.83 \text{ m} = 0.73 \text{ m}$$

Manning's equation for velocity can then be solved and, the discharge computed as

$$V = (1.00/0.013) \bullet 0.73^{2/3} \bullet 0.01^{1/2} = 6.25 \text{ m/s}$$

$$Q = V \bullet A = 6.25 \text{ m/s} \bullet 5.00 \text{ m}^2 = 31.2 \text{ m}^3/\text{s}$$

To answer the second part of the question, we simply repeat the steps for a rectangular section shape.

$$A = (4m \bullet 1m) = 4 \text{ m}^2$$

$$P_w = 4m + 2 \bullet (1m) = 6 \text{ m}$$

$$R = 4 \text{ m}^2 / 6 \text{ m} = 0.67 \text{ m}$$

$$V = (1.00/0.013) \bullet 0.67^{2/3} \bullet 0.01^{1/2} = 5.87 \text{ m/s}$$

$$Q = 5.87 \text{ m/s} \bullet 4 \text{ m}^2 = 23.5 \text{ m}^3/\text{s}$$

As we would expect, this is less than the discharge of the trapezoidal section.

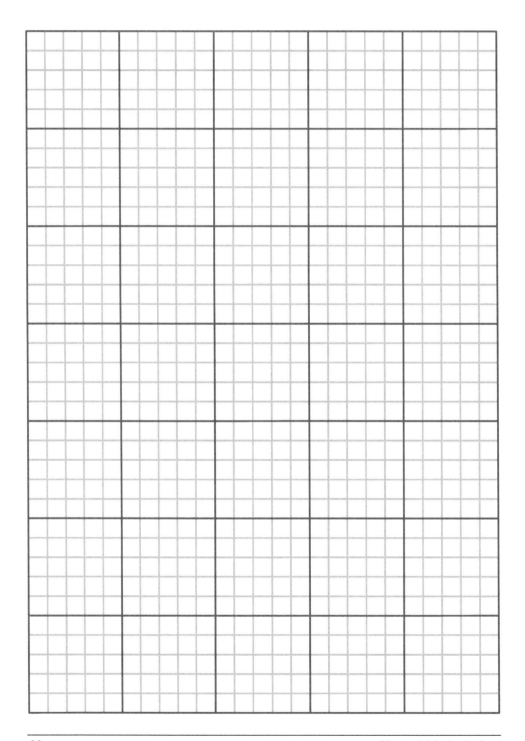

Haestad Methods, Inc.

Specific Energy and Critical Flow

Of course, channels do not always flow at normal depth - if they did, it would make your task as an engineer quite simple. A more in-depth look at non-uniform flow is presented in Chapter 2, but for now let us continue by focusing on another important concept - specific energy. For any flow section, the **specific energy** is defined as the sum of the depth of flow and the velocity head.

$$E = y + \frac{V^2}{2g}$$

where   E   is the specific energy (ft or m)
            y   is the depth of flow (ft or m)
            V   is the mean velocity (ft/s or m/s)
            g   is gravitational acceleration (ft/s^2 or m/s^2)

If we assume the special case of an infinitely short section of open channel (with essentially no friction losses and no change in elevation), we see that the general energy equation can be reduced to an equality of specific energies. In other words,

$$E_1 = y_1 + \frac{V_1^2}{2g} = y_2 + \frac{V_2^2}{2g} = E_2$$

Recall that the velocity of the section is directly related to the area of flow, and the area of flow is a function of channel depth. This means that, for a given discharge, the specific energy at each point is solely a function of channel depth, and there may be more than one depth having the same specific energy. If the channel depth is plotted against specific energy for a given flowrate, the result is similar to the following graph:

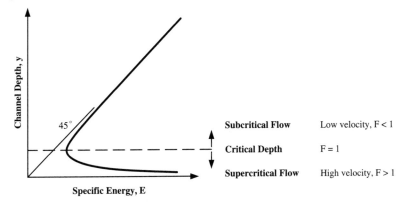

Figure 1-5: Specific Energy

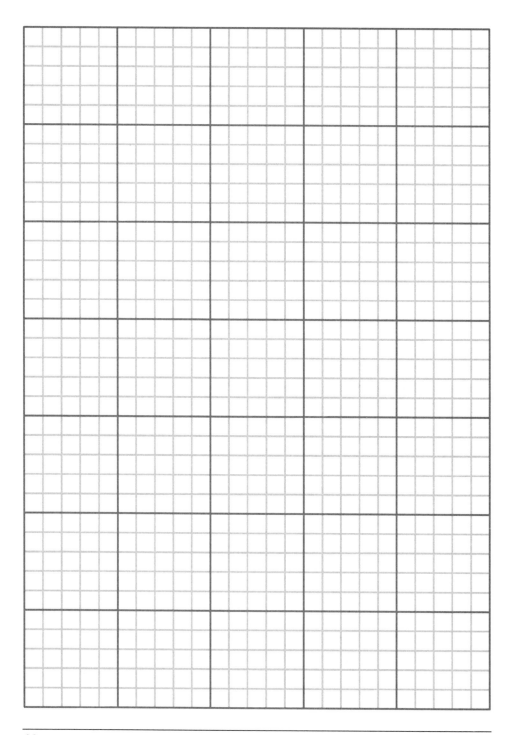

As this figure shows, there is a depth for which the specific energy is at a minimum. This depth is called the **critical** depth. If the velocity is higher than critical velocity (and thus, the depth is below critical depth) the flow is considered to be **supercritical**. If the velocity is lower than critical velocity (and the depth is above critical depth) then the flow is called **subcritical**.

The velocity at critical depth is equal to the wave celerity - the speed at which waves will ripple outward from a pebble tossed into the water (obviously, for open channels only). A unitless value called the **Froude number**, F, represents the ratio of actual fluid velocity to wave celerity. The Froude number is computed as follows:

$$F = \frac{V}{\sqrt{gD}}$$

where   F   is the Froude number
             D   is the hydraulic depth of the channel, defined as A/T
                   A is the flow area (ft^2 or m^2)
                   T is the top width of flow (ft or m)
             V   is the fluid velocity (ft/s or m/s)
             g   is gravitational acceleration (ft/s^2 or m/s^2)

When the flow is at critical depth (velocity is at wave celerity), the Froude number must be equal to one, by definition. This means that the equation can be algebraically rewritten and refactored to form the following equality:

$$\frac{A^3}{T} = \frac{Q^2}{g}$$

where   A   is the flow area (ft^2 or m^2)
             T   is the top width of flow (ft or m)
             Q   is the channel flowrate (ft^3/s or m^3/s)
             g   is gravitational acceleration (ft/s^2 or m/s^2)

This equation can now be used to determine the depth for which this equality holds true - critical depth. While this may be a relatively easy task for simple geometric shapes, it can require quite a bit of iteration before a solution can be found for an irregular shape (such as a natural streambed).

While the determination of critical depth may be a relatively easy task for simple geometric shapes, it can require quite a bit of iteration before a solution can be found for an irregular shape (such as a natural streambed). In fact, for irregular sections there may be several valid critical depths.

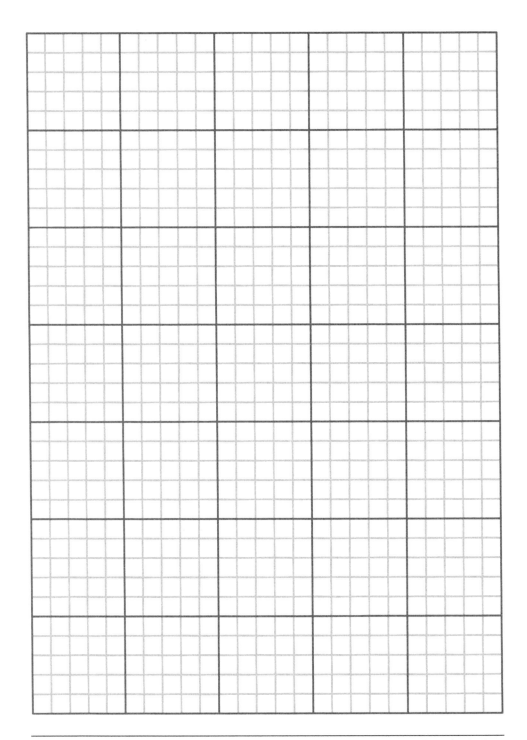

Haestad Methods, Inc.

Example 1-5: Critical Depth

What is the critical depth for a grassy triangular channel with 2H:1V sideslopes and a 0.5% slope when the flow is 3.00 m³/s? If the channel is actually flowing at 1.2 meters deep, is the flow critical, subcritical, or supercritical?

Solution: For such a simple geometry, we can quickly create a relationship between the flow area, top width, and depth of flow:

$$T = 4y \text{ meters}, \quad A = 0.5{\bullet}T{\bullet}y = 0.5{\bullet}4y{\bullet}y = 2y^2 \text{ m}^2$$

Filling these values into the equation for critical depth (above), we can algebraically solve for the channel depth:

$$(2y^2)^3 / 4y = Q^2 / g$$

$$8y^6 / 4y = Q^2 / g$$

$$2y^5 = Q^2 / g$$

$$y^5 = Q^2 / 2g$$

$$y = (Q^2/2g)^{1/5} = [(3.00 \text{ m}^3/s)^2 / (2{\bullet}9.8 \text{ m/s}^2)]^{0.2} = (0.46 \text{ m}^5)^{0.2} = 0.86 \text{ m}$$

The critical depth for this section is 0.86 meters. The actual flow depth of 1.2 meters is above critical depth, so the flow is subcritical.

1.6 Computer Applications

It is very important for students (and practicing engineers) to fully understand the methodologies behind hydraulic computations. Once these concepts are learned, however, the solution process can become repetitive and tedious - the type of procedure that is well suited to computer analysis.

There are several advantages of using computerized solutions for common hydraulic problems:

- The amount of time to perform an analysis can be greatly reduced.

- Computer solutions can be more detailed than hand calculations. Performing a solution manually often requires many simplifying assumptions.

- The solution process may be less error-prone. Unit conversion and the rewriting of equations to solve for any variable are just two examples of where hand calculations tend to have mistakes, while a well-tested computer program avoids these algebraic and numerical errors.

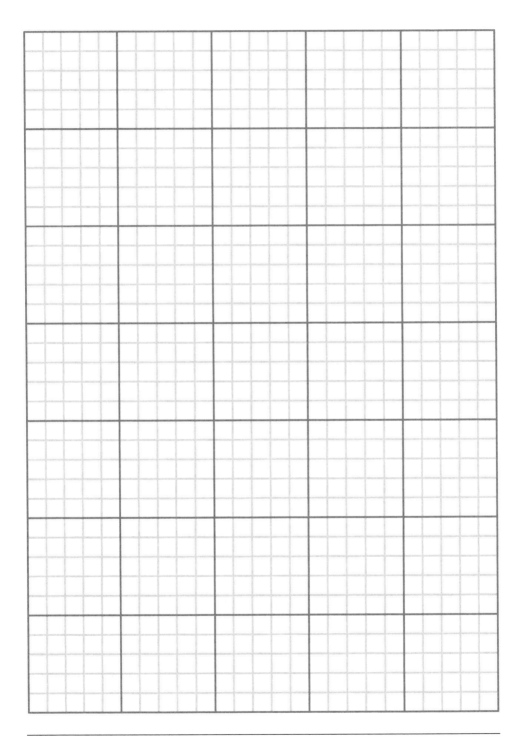

Haestad Methods, Inc.

- The solution is easily documented and reproducible.

- Because of the speed and accuracy of a computer model, more comparisons and design trials can be performed in less time than a single computation would take by hand. This results in the exploration of more design options, which eventually leads to better, more efficient designs.

Take warning, though! Not all software is well designed and bug-free. Try to avoid purchasing software from companies that do not have a proven track record and a reputation for quality. Using software to perform calculations does not relieve the engineer's liability for the accuracy and quality of design.

1.7   Discussion

In order to prevent an "overload" of data, this chapter (and most of this book) deals primarily with steady-state computations. After all, an introduction to hydraulic calculations is tricky enough without throwing in the added complexity of a constantly changing system.

The assumption that a system is under steady-state conditions is oftentimes a perfectly acceptable assumption. Minor changes that occur over time or irregularities in a channel cross-section are frequently negligible, and a more detailed analysis may not be the most efficient or effective use of time and resources.

There are circumstances where an engineer may be called upon to provide a more detailed analysis, though - including unsteady flow computations. For a storm sewer, the flows may rise and fall over time as a storm builds and subsides. For water distribution piping, a pressure wave may travel through the system when a valve is closed abruptly (the same "water-hammer" effect can probably be heard in your house if you close a faucet quickly).

As an engineer, it's important to understand the purpose of an analysis - otherwise you can't possibly choose appropriate methods and tools to meet that purpose.

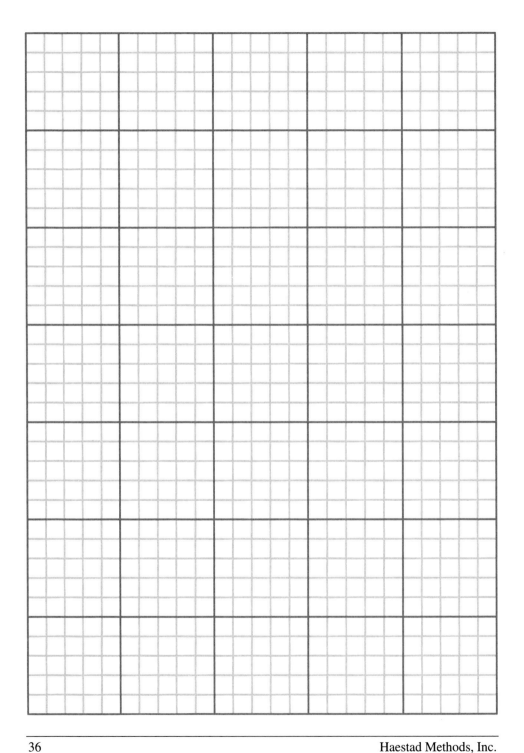

1.8 Problems

Solve the following problems using the FlowMaster computer program (included on the CD that accompanies this textbook). Unless stated otherwise, assume water at 20°C.

1. A rough concrete channel is 6 feet square, on a 0.02 ft/ft slope. Using the Darcy-Weisbach method, determine the maximum allowable flowrate through the channel to maintain one foot of freeboard (freeboard is the vertical distance from the water surface to the overtopping level of the channel). For these conditions, find the following characteristics (note that FlowMaster may not directly report all of these):

   a) flow area

   b) wetted perimeter

   c) hydraulic radius

   d) velocity

   e) Froude number

2. A 450-mm concrete pipe constructed on a 0.6 percent slope carries 0.1 m³/s.

   a) Using Manning's equation and normal depth assumptions, what are the depth and velocity of flow?

   b) What would the velocity and depth be if the pipe were constructed of corrugated metal instead of concrete?

3. A trapezoidal channel carries 2.55 m³/s at a depth of 0.52 meters. The channel has a bottom width of 5 meters, a slope of 1.0 percent, and 2H:1V sideslopes.

   a) What is the appropriate Manning's roughness coefficient?

   b) How deep would the water be if the channel carried 5 m³/s?

4. Use Manning's equation to analyze an existing brick-in-mortar triangular channel with 3H:1V side slopes and a 0.05 longitudinal slope. The channel is intended to carry 7 cubic feet per second during a storm event.

   a) If the maximum depth in the channel is 6 inches, is the existing design acceptable?

   b) What would happen if the channel was replaced by a concrete channel with the same geometry?

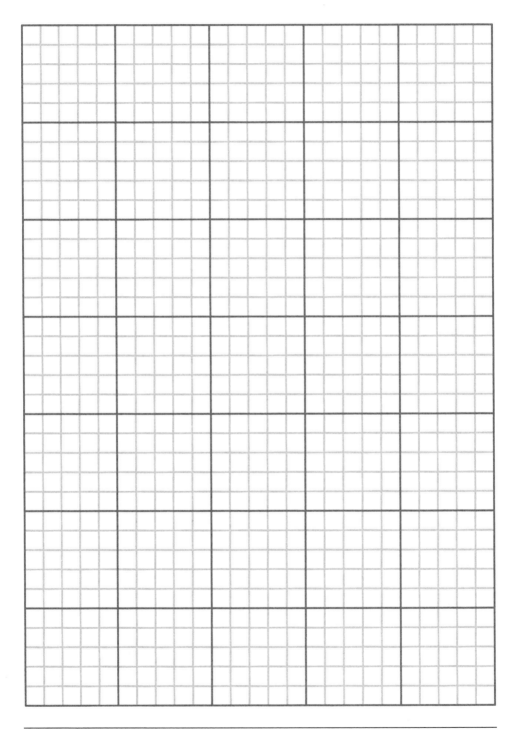

Haestad Methods, Inc.

5. A pipe manufacturer reports that they can achieve Manning's roughness values of 0.011 for their concrete pipes, lower than the 0.013 reported by their competitors. Using Kutter's equation, determine what the difference in flow would be for a 310-mm circular pipe with a slope of 2.5% flowing at one half of the full depth.

6. A grass drainage swale is trapezoidal, with a bottom width of 6 feet, 2H:1V sideslopes. Using whatever friction method you feel is appropriate, answer the following questions:

   a) What is the discharge in the swale if the depth of flow is one foot and the channel slope is 0.005 ft/ft?

   b) What would the discharge be with a slope of 0.010 ft/ft?

7. A paved highway drainage channel has the geometry shown in the following figure. The maximum allowable flow depth is 0.75 feet (to prevent the flow from encroaching on traffic), and the Manning's n value is 0.018 for the type of pavement used.

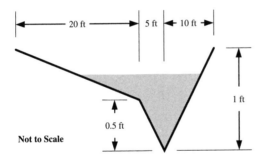

Figure for Problem # 7

   a) What is the capacity of the channel for a 2% longitudinal slope?

   b) Create a rating curve to demonstrate how the capacity varies as the channel slope varies from 0.5% to 5%. Choose a slope interval that will generate a reasonably smooth curve.

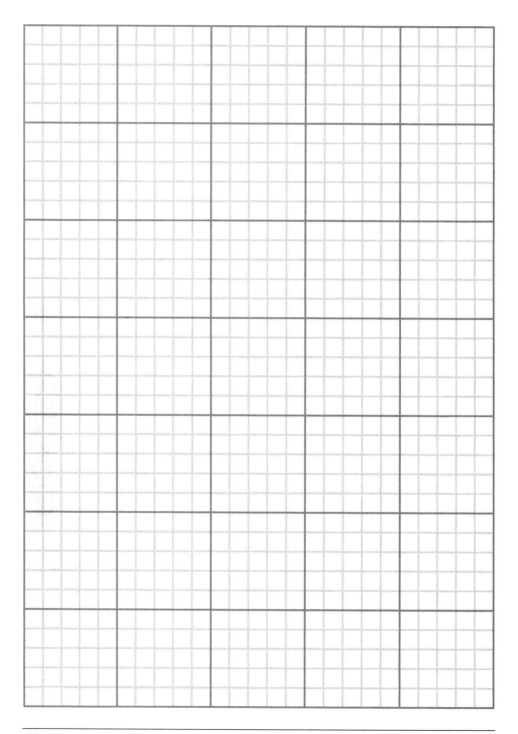

8. Using the Hazen-Williams equation, determine the minimum diameter of a new cast iron pipe for the following conditions: the upstream end is 51.8 meters higher than the downstream end, 2.25 km away. The upstream pressure is 500 kPa, and the desired downstream pressure and flowrate are 420 kPa and 11,000 liters per minute respectively. What minimum diameter is needed? Assume pipes are available in 50-mm increments.

9. 2,000 gallons of water per minute flows through a level 320-yard long, 8-inch diameter cast iron pipe to a large industrial site. If the pressure at the upstream end of the pipe is 64 psi, what will the pressure be at the industry? Is there a significant difference between using the Hazen-Williams method and the Darcy-Weisbach method?

10. Develop a performance curve for the pipe in Problem 9 that shows the available flow to the industry with a residual pressure between 20 psi and 80 psi (assume the source can maintain 64 psi regardless of flowrate). Create similar curves for 10 inch and 12-inch diameter pipes, and compare the differences in flow.

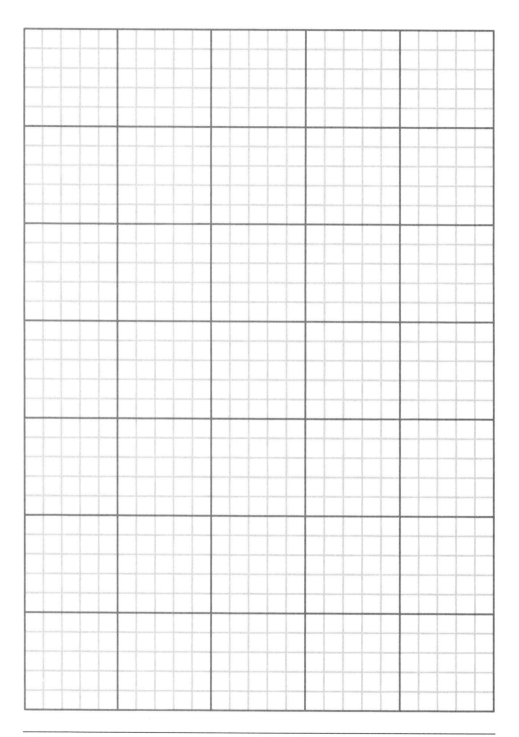

Haestad Methods, Inc.

Chapter 2
Storm Sewer Design

2.1 Hydrologic Principles

Storm sewers are typically designed with one purpose: to carry flow from a rainfall event away from areas where it is unwanted (such as parking lots and roadways). Because of this primary functionality, most storm sewer systems are designed to adequately convey a peak flow rate, based on the characteristics of the watershed and the rainfall event.

To develop a better understanding of the design and analysis of storm sewer systems, it is important to learn a few basics about watershed modeling and hydrology.

Watersheds

A ***watershed*** is an area that drains to a common point of discharge. Since water flows downhill, delineating a watershed is a matter of identifying an outfall, and then locating the boundary such that any rain that falls within the boundary will be directed towards that point of discharge. Because of the collecting nature of river and stream systems, a watershed may have any number of sub-watersheds within it. The focus of analysis (how large does a watershed have to be in order to be analyzed separately) is highly dependent on the scope and purpose of the project at hand.

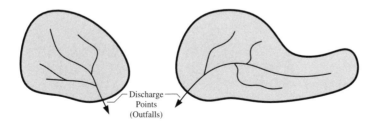

Figure 2-1: Typical Natural Watersheds (with collection channels)

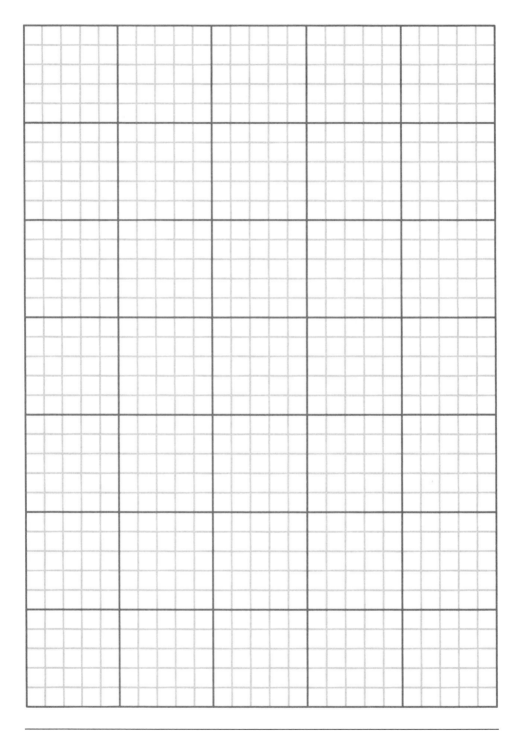

Time of Concentration

The maximum amount of flow coming from any watershed is related to the amount of time it takes for the entire watershed to be contributing to the flow. In other words, it may start raining right now, but it could be several minutes or even hours before the water that lands on some parts of the watershed actually makes its way to the discharge point.

There are some places that are hydraulically closer to the discharge point than others, but for peak flow generation only the most hydraulically remote location is considered crucial. The amount of time that it takes for the first drop of water that lands at this location to work its way to the discharge point is called the *time of concentration*.

There are numerous methods for calculating time of concentration, based on various private, federal, and local publications. Although each of these methods is different (in some cases only slightly) they are all based on the type of ground cover, the slope of the land, and the distance along the flow path. In most locales, there is also a minimum time of concentration (typically 5 to 10 minutes) recommended for small watersheds, like a section of a parking lot draining to a storm sewer.

Return Period and Frequency

Just as there are statistical probabilities of buying a car that falls apart or getting a bad egg at the grocery store, there are also probabilities that a storm event of a certain magnitude will occur in any given year. This hydrological probability is expressed in terms of frequency and the return period.

The *return period* represents, on average, the length of time between rainfall events of a specific magnitude. The *frequency*, or exceedance probability, is a measure of how often a specific rainfall event will be equaled or exceeded, and is simply the inverse of the return period.

For example, a five-year return period represents a storm event that is expected to occur once every five years on the average. This does not mean that two storm events of that size won't occur in the same year, nor does it mean that the next storm event of that size won't occur for another twenty years. It just means that the average will be once every five years.

Intensity and Duration

Rainfall *intensity* is a measure of the rate of rainfall, usually measured in inches per hour or millimeters per hour. The higher the intensity, the "harder" it is raining. The *duration* is indicative of how long the rainfall will fall with that intensity. The longer a rainstorm lasts, the lower the overall intensity will be. This is consistent with the rainfall events

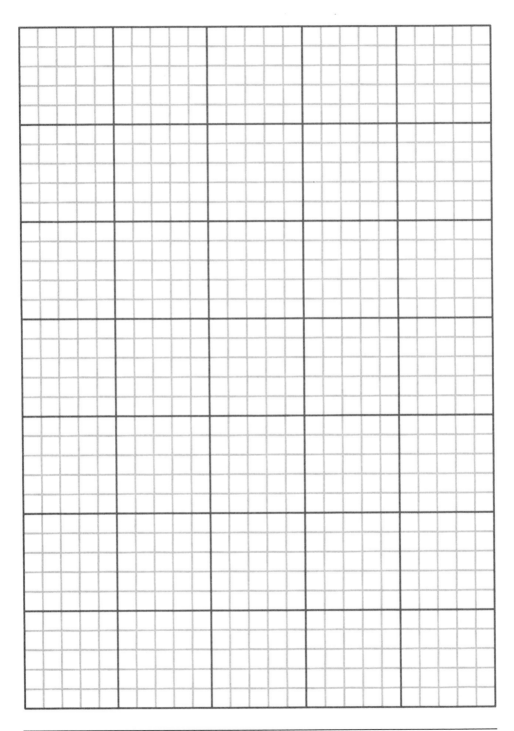

that most of us have witnessed, where it may rain for a day or longer, but there is a period of a few minutes or a few hours when the rain falls harder than at any other time.

2.2 Rational Method Hydrology

The Rational Method is an intensity based rainfall prediction method, meaning that it can be used to predict peak flows based on the characteristics of the watershed and a rainfall event. This makes the Rational Method a popular choice for storm sewer analyses, since it is so well paired with the objectives of these projects.

The basic assumptions of the Rational Method are as follows:

- The drainage area is small

- Peak flow occurs when the entire watershed is contributing

- The rainfall intensity is uniform over a duration of time equal to or greater than the time of concentration

- A rational coefficient relates the ratio of rainfall to runoff, and is considered independent of the rainfall intensity

The Rational Method equation for peak flow is as follows:

$$Q = kCiA$$

where Q is design flow rate (ft^3/s or m^3/s)
  k is a factor to maintain unit consistency;
    1.008 ft^3/s per acre·in/hr for U.S. Standard units, or
    0.00278 m^3/s per hectare·mm/hr for S.I. units
  C is the weighted rational coefficient for the drainage area
  i is the rainfall intensity (in/hr or mm/hr)
  A is the drainage area (acres or hectares)

The Rational Coefficient, C

The rational coefficient, C, is the parameter that is most open to engineering judgment. It is a unitless number between 0 and 1 that relates the rate of rainfall over a watershed to the rate of discharge from that watershed. The coefficient is highly dependent on the land use and slope, as can be seen in the following table. The more covered and impervious the soil is (such as pavement) the closer to 1 the C value is.

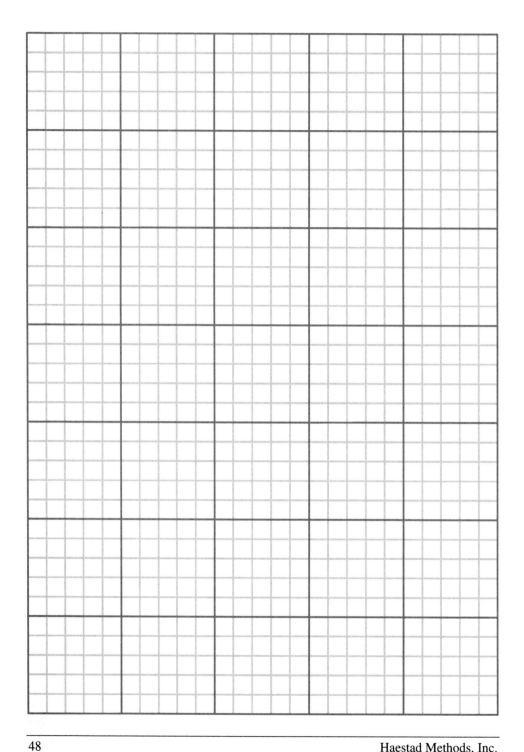

Area Type	Description	C Values
Business	Downtown	0.70-0.95
	Neighborhood	0.50-0.70
Residential	Single Family	0.30-0.50
	Multiunit detached	0.40-0.60
	Multiunit attached	0.60-0.75
	Suburban resident	0.25-0.40
	Apartment	0.50-0.70
	1.2 acre lots or more	0.30-0.45
Lawns, Heavy Soils	Flat 2%	0.13-0.17
	Average, 2-7%	0.18-0.22
	Steep >7%	0.25-0.35
Industrial	Light	0.50-0.80
	Heavy	0.60-0.90
Lawns, Sandy Soils	Flat, 2%	0.05-0.10
	Average, 2-7%	0.10-0.15
	Steep > 7%	0.15-0.20
Pavement	Asphalt / Concrete	0.70-0.95
	Brick	0.70-0.85
	Drives and Walks	0.75-0.85
Miscellaneous	Parks and Cemeteries	0.10-0.25
	Playgrounds	0.20-0.40
	Railroad Yards	0.10-0.30
	Unimproved	0.20-0.40

Table 2-1: Typical Rational C Coefficients

Of course, most watersheds are comprised of more than one type of land cover. For example, a storm sewer inlet may accept flow from a paved area, a grassy area, and a wooded area. To composite these values, all the engineer needs to do is multiply each corresponding C and A pair, and sum the values to obtain the total C·A (or CA) for the entire watershed. Note that since C is unitless, CA has the same units as the area.

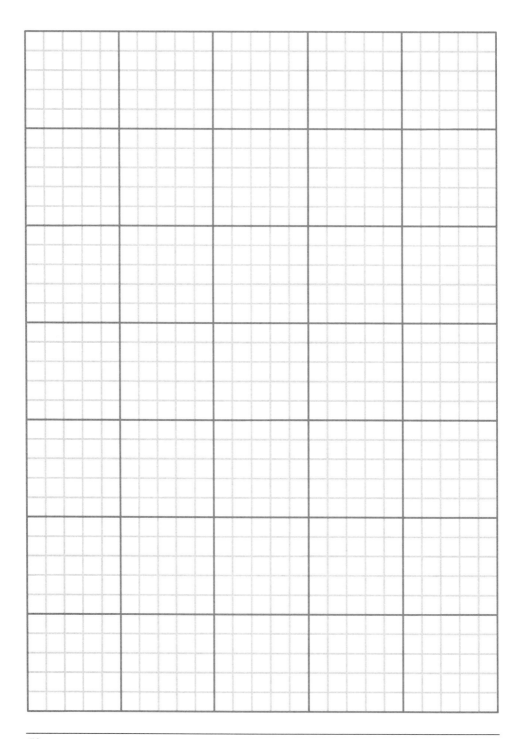

Haestad Methods, Inc.

Example 2-1: Weighted CA for a Watershed

A single watershed contains 0.50 acres of pavement, 0.25 acres of steep lawn, and 0.90 acres of trees and underbrush. What is the weighted CA value for this watershed?

Solution:

Total CA = 0.50ac·0.90 + 0.25ac·0.30 + 0.90ac·0.20 = 0.70 acres

Intensity-Duration-Frequency (IDF) Curves

As mentioned previously, the intensity of a storm is directly related to the duration of the storm, and the return period (and frequency) of the storm event. Historical storm data is compiled and analyzed to predict the storm characteristics, and is often presented in the form of Intensity-Duration-Frequency (IDF) curves. These curves are usually available from a local regulatory agency or weather bureau. An example set of IDF curves are presented in the following figure:

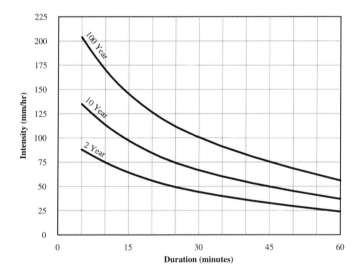

Figure 2-2: Example IDF Curves

Rainfall Tables and Equations

While graphical rainfall curves may be acceptable for hand calculations, they are not well suited to computer analysis. This is why software programs, including StormCAD, prefer the data in tabular format (so the model can fit a curve to the data and interpolate intermediate points), or in an equation form.

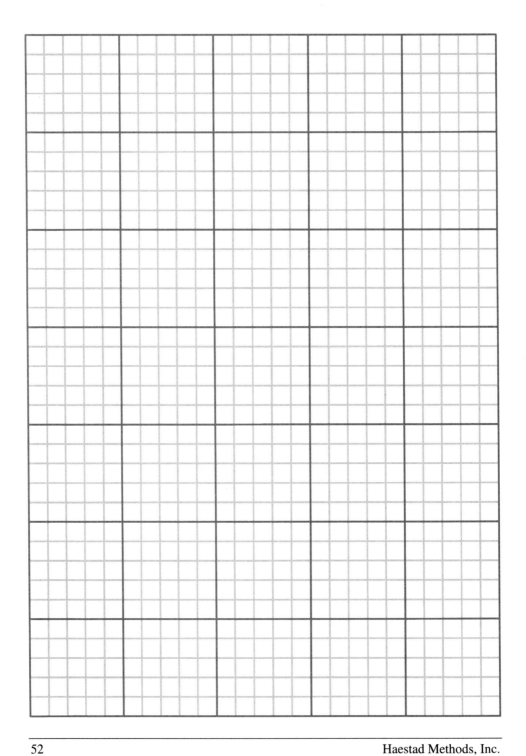

Creating a rainfall table is a simple matter of manually picking values off a set of rainfall curves, and entering them into the table. For example, the following table could be created from the data presented in the previous IDF curves:

Rainfall Intensities (mm/hr)			
Storm	Return Period		
Durations	2 Year	10 Year	100 Year
5 min	88	135	204
10 min	75	114	168
15 min	65	97	142
30 min	44	66	100
60 min	24	37	56

Table 2-2: Example IDF Table

Choosing equation coefficients is a much more substantial task than the creation of a rainfall table. These equations are typically used only in regions where the rainfall data has already been analyzed and an appropriate equation has been fit to the data. The most common forms of these equations are:

$$i = \frac{a}{(b+D)^n} \qquad i = \frac{a(R_P)^m}{(b+D)^n} \qquad i = a + b(\ln D) + c(\ln D)^2 + d(\ln D)^3$$

where i is intensity of rainfall (usually in/hr or mm/hr)
D is the rainfall duration (usually min or hrs)
R_P is the return period (usually in years)
$a, b, c, d, m,$ and n are coefficients

2.3 Gradually Varied Flow

In some areas of hydraulic analysis, the change in depth along the length of the section may be so gradual that the entire section can be assumed to have a constant depth (such as normal depth) without any loss of accuracy. In other systems, however, such as storm sewers, there may be some sort of restriction that prevents the flow depth from equaling normal depth throughout the length. For example, a high tailwater elevation may force the depth to be above normal depth at the downstream end of a pipe, as shown in the following figure.

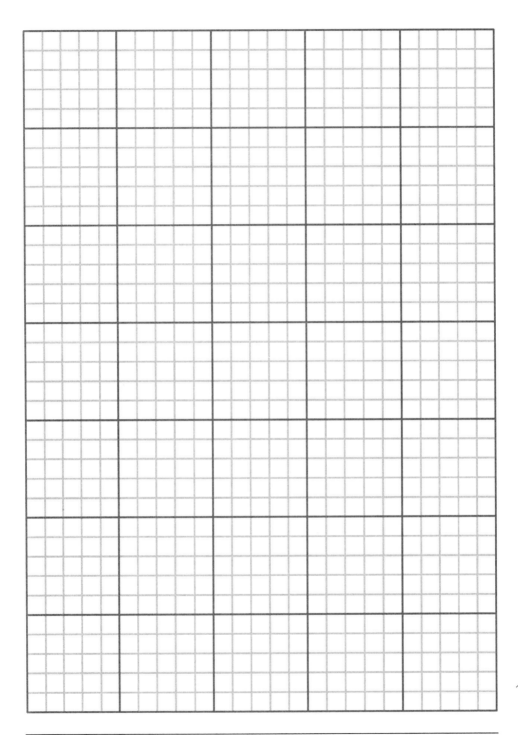

Haestad Methods, Inc.

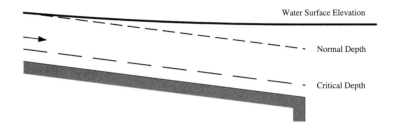

Figure 2-3: Non-Uniform Flow in an Open Channel

When these differences in depth from one end of a conduit to the other are significant, a more detailed form of analysis is required, called gradually varied flow. Gradually varied flow analysis is the process of splitting a channel length into smaller segments, and analyzing each segment separately. It is based on several assumptions, including the following:

- The head loss within any given calculation segment is the same as for uniform flow conditions

- The velocity is the same across the entire cross-section

- The slope of the conduit is less than 10%

- The roughness coefficient is constant throughout the reach under consideration and is independent of the depth of flow

- The depth of flow changes gradually along the length of the conduit, starting from some controlling boundary condition (usually the tailwater elevation); there are no sudden increases or decreases in depth

- If the pipe is sufficiently long enough, the depth of flow will approach normal depth

Flow Classification

The first step in performing a gradually varied flow analysis is to identify the type of flow that is expected to occur in the conduit, based on the slope of the channel, normal depth, critical depth, and the controlling boundary.

Slope Classification

Once the normal depth and critical depth have been computed for the section, the conduit slope can be determined. If the normal depth is above critical depth, the slope is said to be *mild*. If the normal depth is equal to the critical depth, the slope is said to be *critical*. Likewise, if normal depth is below critical depth, then the slope is said to be *steep*.

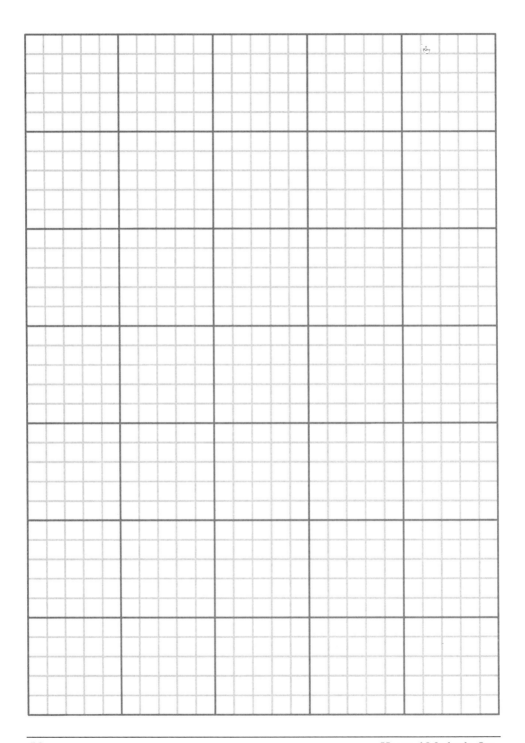

There are also two other types of slope to consider, for which there is no normal depth defined. One type is a *horizontal* channel, and the other is an *adverse* slope (an "uphill" sloping channel).

For a gradually varied flow profile, the first letter of the flow type is used in the identification of the profile. For example, a channel with a hydraulically steep slope is a type S, a channel with a hydraulically mild slope is a type M, and so forth.

Flow Zone Classification

In a gradually varied flow analysis, the flow is assumed to stay within the same zone throughout the length of the conduit. For example, the flow will not jump from subcritical to supercritical within the same profile type. There are three zones where flow profiles can occur:

- Zone 1: Actual flow depth is above both normal and critical depth

- Zone 2: Actual flow depth is between normal depth and critical depth

- Zone 3: Actual flow depth is below both normal and critical depth

A given profile will exist in only one of these zones. Because normal depth is undefined for infinite for horizontal slopes and undefined for adverse slopes, zone one flow does not exist for profiles with these slopes.

Profile Classifications

Once the slope and flow zone have each been classified, the profile type can be defined, and the engineer can determine how to proceed with the hydraulic grade computations. The figure on the following page shows the basic profile types.

For computation, the engineer must determine from the profile type whether the flow is subcritical or supercritical (based on the location of actual depths compared to critical depth). In order to prevent excessive velocities that could cause pipe scour or channel erosion, most storm sewers are designed with mild slopes to carry subcritical flows. This means that the hydraulic control is at the downstream end of the section, and proceeds towards the upstream end. When the flow depth is above normal depth (as in an M1 profile), this type of analysis is called a *backwater* analysis. When the flow depth is between critical depth and normal depth, it is called a *drawdown* analysis.

When supercritical flows are encountered, the controlling section is at the upstream end of the conduit and the computations proceed from upstream to downstream. We will consider this calculation type to be a *frontwater* analysis.

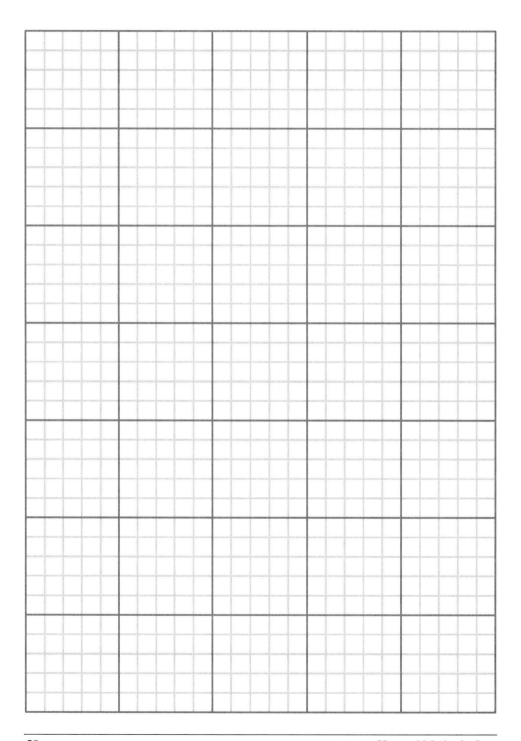

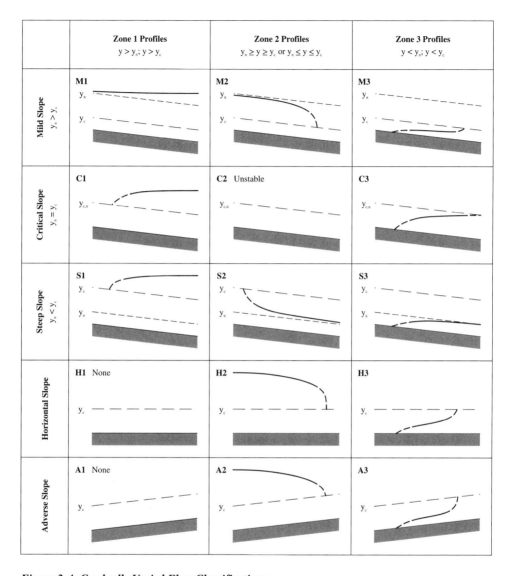

Figure 2-4: Gradually Varied Flow Classifications

Energy Balance

Even for gradually varied flow, the solution is still a matter of balancing the energy between the two ends of the segment. The energy equation as it relates to each end of a

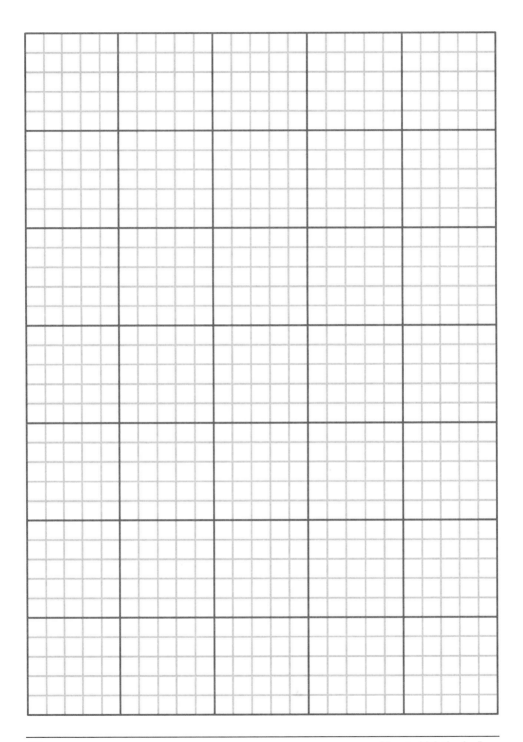

Haestad Methods, Inc.

segment is as follows (note that the pressures for both ends are zero, since it is free surface flow):

$$Z_1 + \frac{V_1^2}{2g} = Z_2 + \frac{V_2^2}{2g} + H_L$$

where Z_1 is the hydraulic grade at the upstream end of the segment
              V_1 is the velocity at the upstream end
              Z_2 is the hydraulic grade at the downstream end of the segment
              V_2 is the velocity at the downstream end
              H_L is the loss due to friction (other losses are assumed to be zero)
              g is gravitational acceleration

The friction loss is computed based on the average rate of friction loss along the segment, and the length of the segment. This relationship is as follows:

$$H_L = S_{Avg} \cdot \Delta x = \frac{S_1 + S_2}{2} \Delta x$$

where H_L is the loss across the segment
              S_{Avg} is the average friction loss
              S_1 is the friction slope at the upstream end of the segment
              S_2 is the friction slope at the downstream end of the segment
              Δx is the length of the segment being analyzed

The conditions at one end of the segment are known (through assumption or from a previous calculation step). Since the friction slope is a function of velocity, which is a function of depth, the depth at the other end of the segment can be found through iteration. There are two primary methods for this iterative solution, the Standard Step method and the Direct Step method.

Standard Step Method

This method involves dividing the channel into segments of known length, and solving for the unknown depth at one end of the segment (starting with a known or assumed depth at the other end). The standard step method is the most popular method of determining the flow profile because it can be applied to any channel, not just prismatic channels.

Direct Step Method

The direct step method is based on the same basic energy principles as the standard step method, but takes a slightly different approach towards the solution. Instead of assuming a segment length and solving for the depth at the end of the segment, the direct step method assumes a depth and then solves for the segment length.

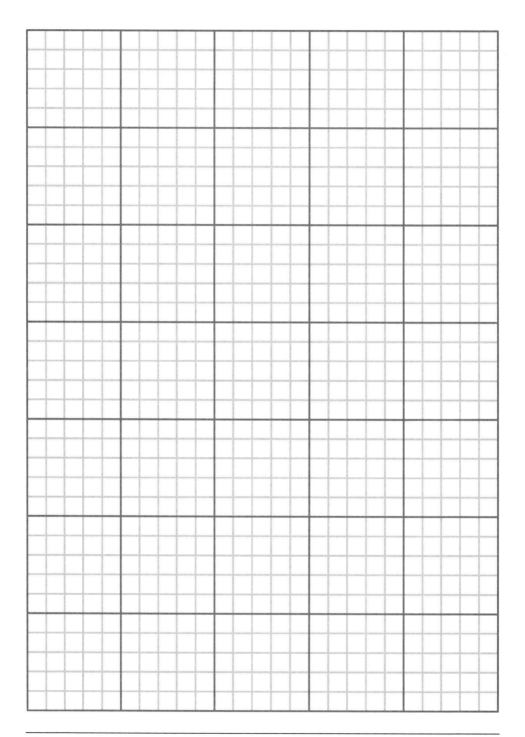

Haestad Methods, Inc.

2.4 Mixed Flow Profiles

So far, we have only looked at conditions where the entire channel has the same profile type, which results in the smooth curves of a gradually varied flow analysis. This section explores those cases where profile types are mixed within the same section, and the steps that can be taken to analyze these occurrences.

Sealing Conditions

There may be conditions such that part of the section is flowing flow, while part of the flow remains open. These conditions are called *sealing* conditions, and are analyzed in separate parts. The portion of the section flowing full is analyzed as pressure flow, and the remaining portion is analyzed with gradually varied flow.

Rapidly Varied Flow

Rapidly varied flow is turbulent flow resulting from the abrupt and pronounced curvature of flow streamlines into or out of a hydraulic control structure. Examples of rapidly varied flow include hydraulic jumps, bends and bridge contractions.

Hydraulic Jumps

When flow passes rapidly from supercritical to subcritical flow, a hydraulic phenomenon called a *hydraulic jump* occurs. The most common occurrence of this within storm sewer networks occurs when there is a steep pipe discharging into a particularly high tailwater, as shown in the following figure.

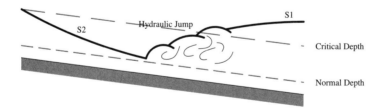

Figure 2-4 Plot of Hydraulic Jump

There are significant losses associated with hydraulic jumps, due to the amount of mixing and turbulence that occurs. These forces are also highly erosive, so engineers typically try to prevent jumps from occurring in storm sewer systems, or at least try to predict the location of these jumps in order to provide adequate channel, pipe, or structure protection.

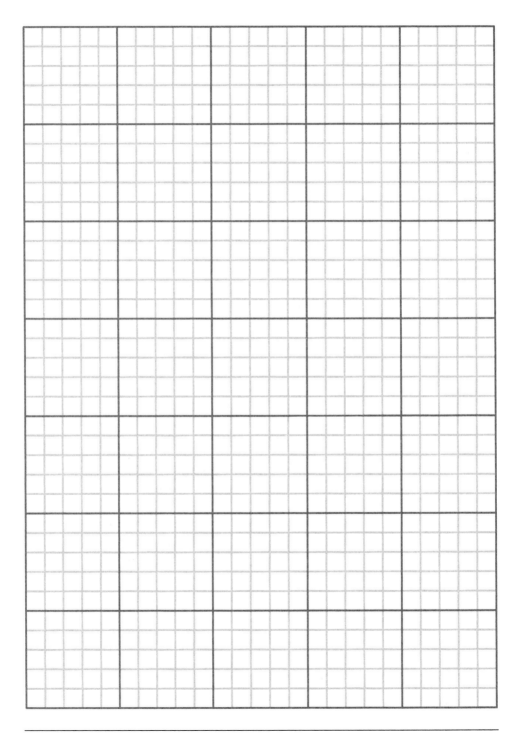

Haestad Methods, Inc.

2.5 Storm Sewer Applications

Storm sewer analysis occurs in two basic calculation sequences:

- Hydrology - Watersheds are analyzed and flows are accumulated from upstream inlets towards the system outlet

- Hydraulics - An assumed tailwater condition at the outlet and the flow values (from the hydrology calculations) are used to compute hydraulic grades from the outlet towards the upstream inlets

Hydrology Model

As the runoff from a storm event travels through a storm sewer, it combines with other flows and the resulting flows are based on the overall watershed characteristics. As with a single watershed, the peak flow is assumed to occur when all parts of the watershed are contributing to the flow, so the rainfall intensity is based on the controlling system time:

- Local time of concentration

- Upstream system time (including pipe travel times)

Whichever of these times is larger is the controlling system time, and is used for computing the intensity (and therefore the flow) in the combined system.

Example 2-2: Flow Accumulation

A storm sewer inlet has a local time of concentration of 8 minutes for a watershed with a weighted CA value of 1.23 acres. This inlet discharges through a pipe to another storm sewer inlet with a weighted CA of 0.84 acres and a local time of concentration of 9 minutes. If the travel time in the pipe is 2 minutes, what is the overall system CA and corresponding storm duration?

Solution: The total CA can be found by simply summing the CA values from the two inlets. The storm duration, however, must be found by comparing the time of concentration at the second inlet to the total time for flow from the upstream inlet to reach the downstream inlet.

> Total CA = 1.23 acres + 0.84 acres = 2.07 acres
>
> Upstream Time = 8 minutes + 2 minutes = 10 minutes

The total upstream flow time of 10 minutes is greater than the local time of concentration at the downstream inlet (9 minutes). The 10 minutes value is therefore the controlling time, and should be used as the duration of the storm event.

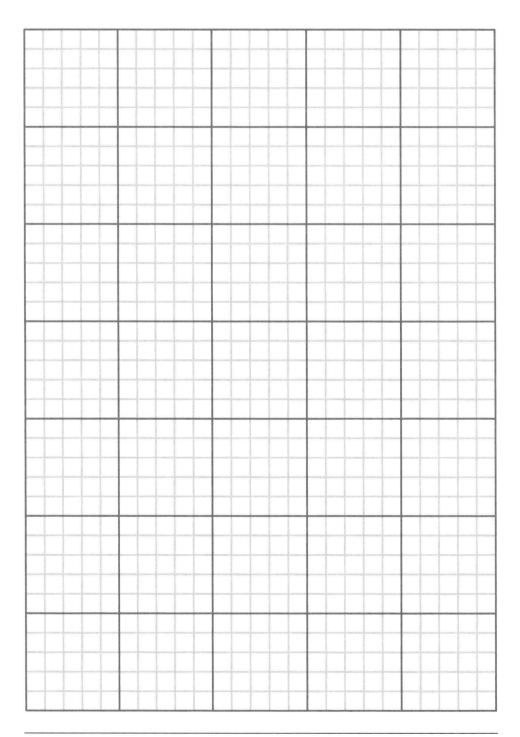

Haestad Methods, Inc.

Other Sources of Water

Other than direct watershed inflow, there may be a number of other flow sources for a storm sewer system. Flow may be piped into an inlet from an external connection, or there could be flow entering an inlet that is carryover (bypass) from another storm sewer inlet.

Design practices vary from jurisdiction to jurisdiction as far as these carryover flows are concerned. Some local regulations may require that pipes be sized to include all flow that arrives at an inlet, while other locales may specify that pipes be sized to accommodate only the flow that is actually intercepted by each inlet. It is the design engineer's responsibility to ensure that the design is in agreement with whatever the local criteria and policies may be.

Hydraulic Model

The hydraulics of a storm sewer are often computed as described previously in this chapter, and may have a combination of sections that flow under normal depth conditions, pressure (submerged) conditions, and both gradually and rapidly varied flow conditions.

The place where all of the computations start is at the system outlet , where a tailwater condition must be assumed. There are four basic assumptions for tailwater conditions:

- **Normal Depth**--The depth at the outfall of the farthest downstream pipe is assumed to be at normal depth (as for a sufficiently long S2 profile)

- **Critical Depth**--The depth at the pipe outfall is assumed to be critical depth, as in subcritical flow to a free discharge

- **Crown Elevation**--The depth is set to the crown (top) of the pipe for free outfall

- **User-specified Tailwater**--A fixed tailwater depth can also be used, as when there is a known pond or river water surface elevation at the outfall of the storm sewer

Care should be taken to choose an accurate tailwater condition, since this will affect the rest of the system.

The designer of a storm sewer system should consider the tailwater depth during storm conditions. An outlet may be above the receiving stream during dry weather, but can be submerged during the design storm event.

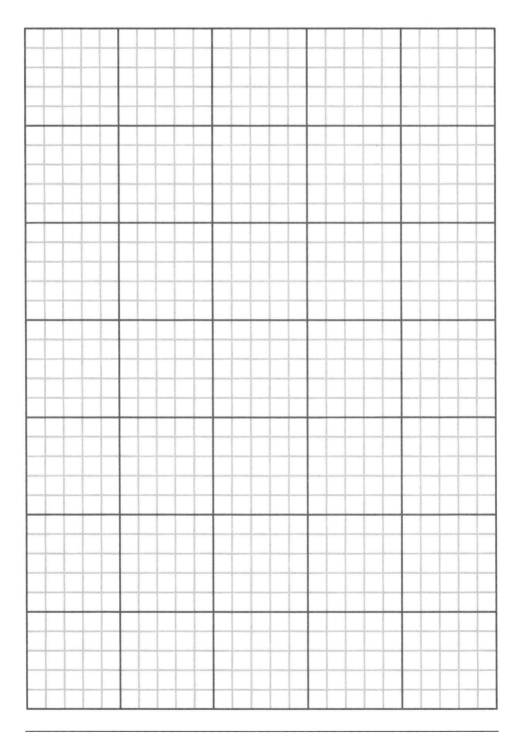

2.6 Problems

Solve the following problems using the StormCAD computer program developed by Haestad Methods (included on the CD accompanying this text).

1. The following data describes an existing storm sewer system shown. For runoff calculations, assume C = 0.3 for pervious cover and C=0.9 for impervious land cover. The ground elevation at the system discharge point is 17.0 m. All pipes are concrete (n=0.13).

   a) Analyze the system for a design return period of 10 years. Assume a free outfall condition. Provide output tables summarizing pipe flow conditions and hydraulic grades in the inlets. How is this system performing?

   b) Increase the size of pipe P-3 to 450-mm. Re-run the analysis and present the results. How does the system perform with this improvement?

   c) Local design regulations require that storm sewer systems handle 25-year return periods without flooding. Re-run the analysis for the improved system in (b). Does the system meet this performance requirement?

   d) The above analyses are run using the default program Manning's n of 0.013. Many drainage design manuals propose a less conservative design roughness of 0.012. Re-analyze the improved system under 25 year flows using n=0.012. How does this influence the predicted performance of the system?

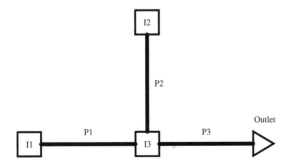

Schematic for Problem 1

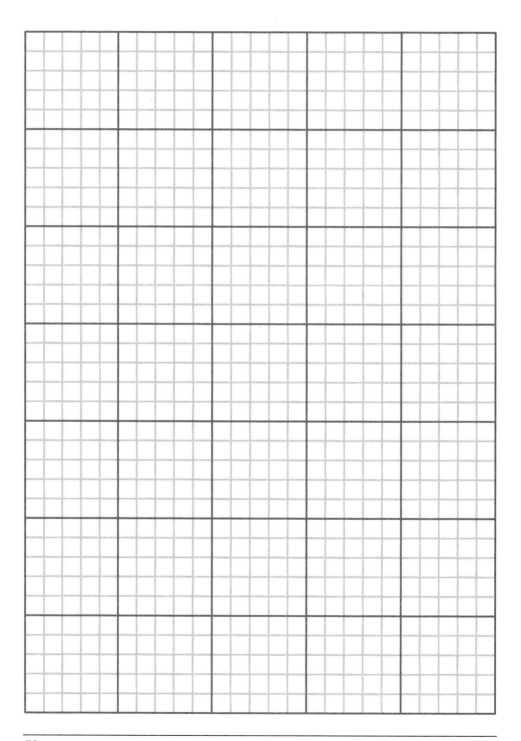

Duration	Rainfall Intensity (mm/hr)		
(min)	5 year	10 year	25 year
5	165	181	205
10	142	156	178
15	123	135	154
30	91	103	120
60	61	70	80

Rainfall Data for Problem 1

Inlet	Ground Elevation (m)	Impervious Area (ha)	Pervious Area (ha)	Time of Concentration (min)
I-1	17.9	0.13	0.32	6.0
I-2	18.0	0.15	0.58	5.0
I-3	17.6	0.08	0.36	5.0

Inlet Information for Problem 1

Pipe	Upstream Invert (m)	Downstream Invert (m)	Diameter (mm)	Length (m)
P-1	16.7	16.15	300	56
P-2	16.8	16.1	375	46
P-3	16.1	15.3	375	54

Pipe Information for Problem 1

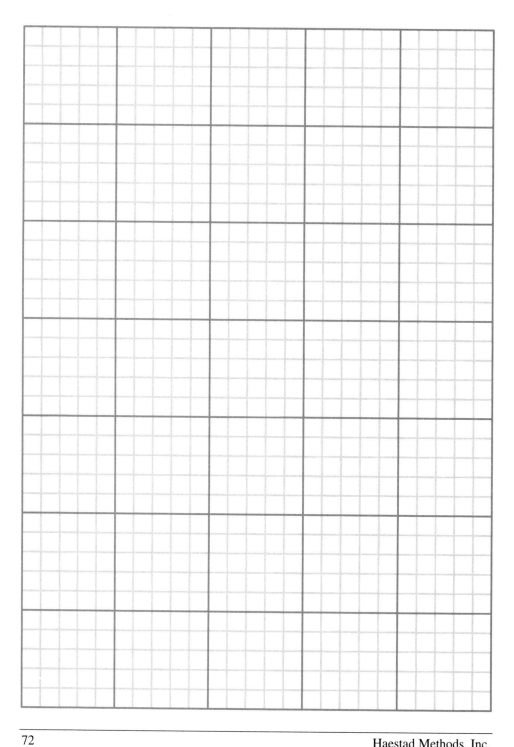

Haestad Methods, Inc.

2. You have been asked by the lead project engineer for a water supply utility to design the stormwater collection system for the proposed ground storage tank and pump station facility shown in the layout. Pipe lengths for P-1, P-2, P-3, and P-4 are 88, 92, 185, and 46 feet respectively.

a) Using the StormCAD program's "Automated Design" feature, size the system using the design data provided below. Use concrete pipe (n=0.013). Use the 25-year S.I. intensity duration frequency data provided in problem 1. (Hint: StormCAD can mix S.I. and U.S. Standard units). The top of bank elevation at the outfall ditch is 840.0 ft. The outfall pipe invert must be located at or above elevation 835.0 ft. Assume the water surface elevation at the outfall is 839± feet. Present your design in tabulated form, and provide a profile plot of your design.

b) During agency review, the county engineer requests that the water utility and the county work cooperatively to accommodate the planned construction of an elementary school nearby, by increasing the size of the proposed storm system so that it can handle the design runoff from the school. The county engineer performs his own calculations and asks that you increase the size of pipes P-3 and P-4 to handle an additional contributing area, CA, of 9.5 acres with a time of concentration of 8.00 minutes. Using StormCAD, introduce the additional flow at inlet I-3 and revise the facility design using the Automated Design functionality of the program. Are the design velocity constraints met? What can you say about the flow conditions in Pipe P-3 and Pipe P-4?

c) Manually fine-tune and revise the design to meet all design criteria. Document your design as in part a).

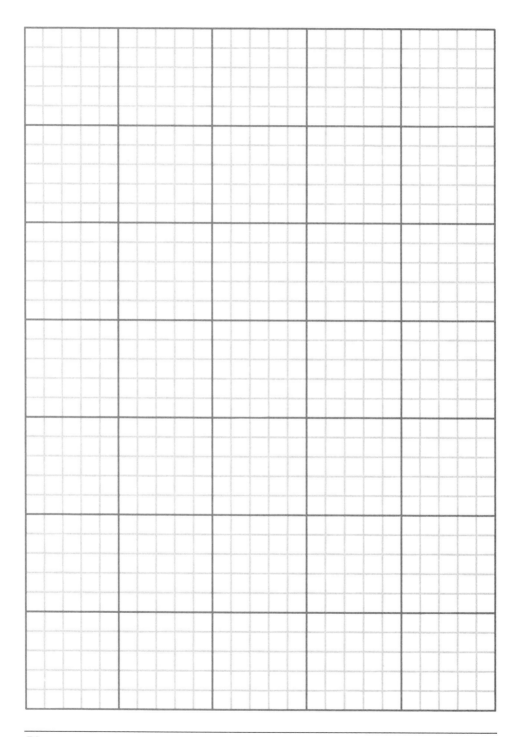

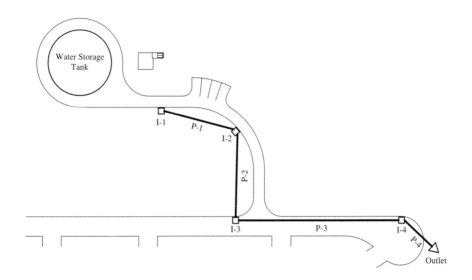

CAD Drawing for Problem 2

Inlet	Ground Elevation (ft)	Impervious Area (ac)	Pervious Area (ac)	Time of Concentration (min)
I-1	865.2	0.25	0.25	6.0
I-2	862.4	0.09	-	5.0
I-3	856.3	0.20	0.49	5.0
I-4	845.8	0.18	0.5	7.5

Inlet Information for Problem 2

Constraint	Minimum	Maximum
Velocity	2 ft/s	15 ft/s
Cover	4 ft*	--
Pipe Slope	0.005 ft/ft	0.100 ft/ft

Design Constraints for Problem 2

*Cover constraints may be relaxed at the outfall, where invert locational requirements will typically govern.

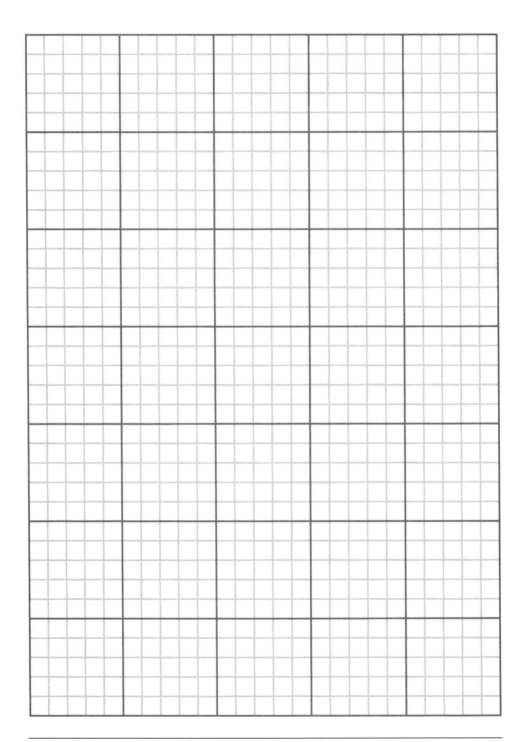

Haestad Methods, Inc.

Chapter 3
Culvert Hydraulics

3.1 Culvert Systems

Culverts are commonplace in practical hydraulic design, including applications such as roadway crossings and detention pond outlets. A roadway cross drainage culvert is typically designed to carry flows from one side of the road to another, without allowing the headwater (water surface elevation just upstream from the culvert) to exceed safe levels. When engineers analyze these culvert systems, they are usually trying to solve for one or more of the following:

- The size, shape and number of new or additional culverts required to pass a design discharge

- The hydraulic capacity of an existing culvert system under an allowable headwater constraint

- The upstream flood level at an existing culvert system resulting from a discharge rate of special interest;

- Hydraulic performance curves for a culvert system, to assess hydraulic risk at a crossing or to use as input to another hydraulic or hydrologic model.

Similar to a storm sewer system, a culvert system consists of a hydrology component, a culvert component, and a tailwater component. While the watershed hydrology and tailwater conditions are almost identical to those that you would normally analyze for a storm sewer or other open channel transport system, there are additional hydraulic computations that are typically reserved for culvert analysis only.

Culvert Hydraulics

Obtaining accurate solutions for culvert hydraulics can be a formidable computational task. Culverts act as a significant constriction to flow and are subject to a range of flow types including both gradually varied and rapidly varied flow.

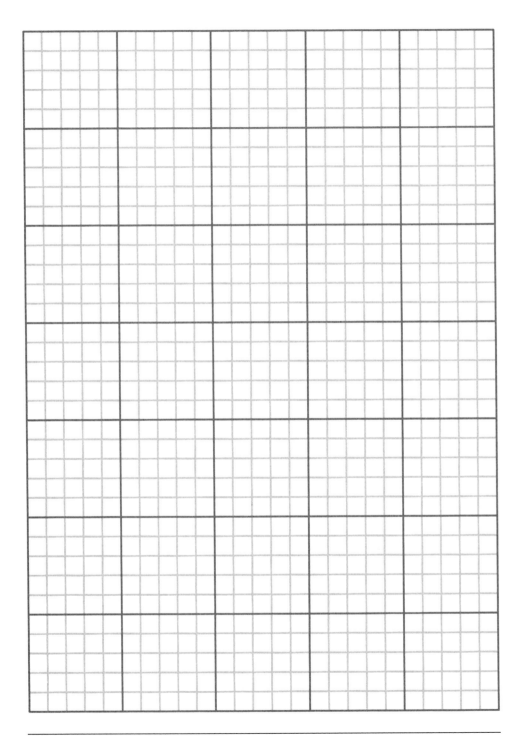

Haestad Methods, Inc.

It is this mix of flow conditions and the highly transitional nature of culvert hydraulics that make the hydraulic solutions so difficult. For this reason, the documented approach is to simplify the hydraulics problem and analyze the culvert system using two different assumptions of flow control:

- **Outlet control** assumption - Computes the upstream headwater depth using conventional hydraulic methodologies that considers the predominant losses due to the culvert barrel friction as well as the minor entrance and exit losses; the effect of the tailwater condition during the design storm has an important affect on the culvert system

- **Inlet control** assumption - Computes the upstream headwater depth resulting from the constriction at the culvert entrance while neglecting the culvert barrel friction and other minor losses

The controlling headwater depth is the larger value of the computed inlet control and outlet control headwater depths. Because the culvert system may operate under inlet control conditions for a range of flow rates and under outlet control conditions for another range of discharges, calculations must be performed for both control conditions every time.

3.2 Outlet Control Hydraulics

Outlet control headwater depths are computed similarly to storm sewer analysis. The headwater depth is found by summing the entrance minor loss, exit minor loss, and friction losses along the culvert barrel. The energy basis for solving the outlet control headwater (HW) for a culvert is presented graphically in the figure and the basic energy equation.

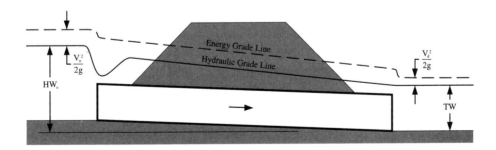

Figure 3-1: Full Flow Characteristics

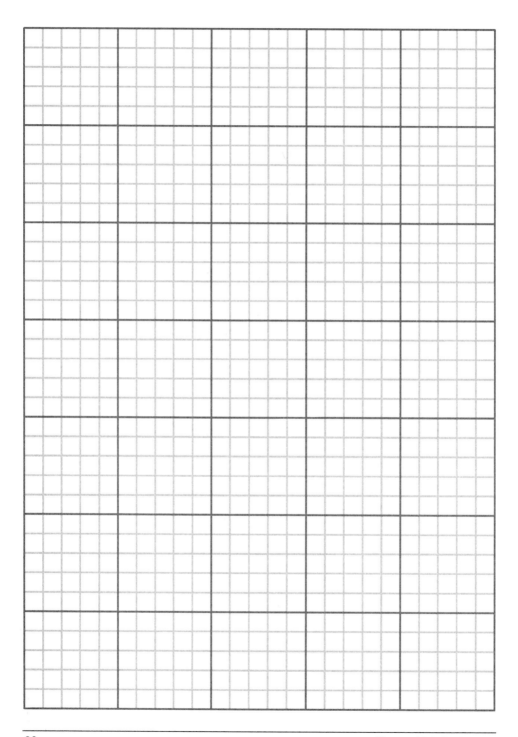

Notice that the energy grade line and hydraulic grade line are parallel throughout the length of the barrel, since full flow results in the same velocity (and therefore the same velocity head) throughout the length of the culvert. At the culvert entrance, there is a slight increase in velocity and dip in the HGL, caused by the flow contraction that occurs there.

The energy equation can be rewritten specifically for culvert terms, which results in the following form:

$$HW_o + \frac{V_u^2}{2g} = TW + \frac{V_d^2}{2g} + H_L$$

where HW_o is the headwater depth above the outlet invert (ft or m)
          V_u is the approach velocity (ft/s or m/s)
          TW is the tailwater depth above the outlet invert (ft or m)
          V_d is the exit velocity (ft/s or m/s)
          H_L is the sum of all losses, including the entrance minor loss (H_E), barrel friction losses (H_F), the exit loss (H_O), and other losses (ft or m)

Culverts often discharge between ponds or other bodies with negligible velocities, so the approach velocity and the velocity downstream of the culvert are often neglected, resulting in the following equation:

$$HW_o = TW + H_L$$

where HW_o is the headwater depth above the outlet invert (ft or m)
          TW is the tailwater depth above the outlet invert (ft or m)
          H_L is the sum of all losses as listed above (ft or m)

Friction Losses

Culverts are frequently hydraulically short, and uniform flow depths are not always achieved. For this reason, gradually varied flow methods are well suited to the analysis. For a more detailed description of gradually varied flow, see Chapter 2.

Entrance Minor Loss

The entrance loss is caused by the contraction of flow as it enters the culvert, and is a function of the barrel velocity head just inside the entrance. It is expressed by the following equation:

$$H_e = k_e \left(\frac{V^2}{2g} \right)$$

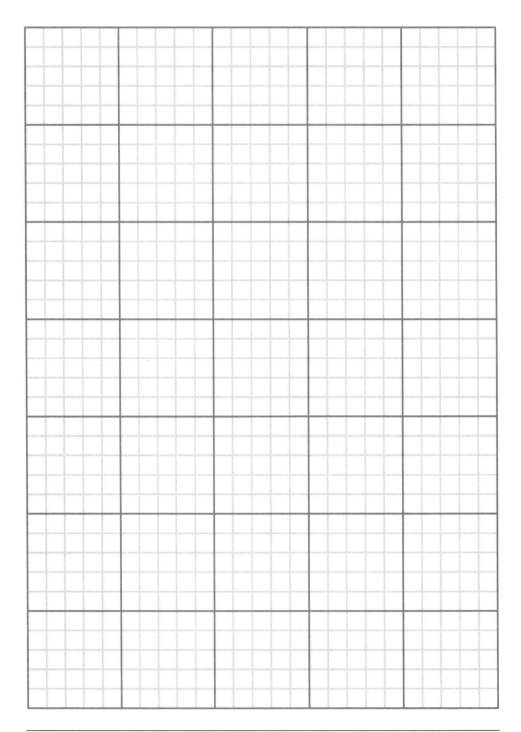

Haestad Methods, Inc.

where k_e is the entrance loss coefficient

         V is the velocity just inside the barrel entrance (ft/s or m/s)

         g is gravitational acceleration (ft/s^2 or m/s^2)

The entrance loss coefficient varies depending on the type of inlet that is present. The smoother the transition is from the channel or pond into the culvert, the lower the loss coefficient is. Values for the coefficients are presented in the following table.

Culvert Type	Entrance Type and Description	Entrance Loss Coefficient, k_e
Pipe, Concrete	Projecting from fill, socket end (groove-end)	0.2
	Projecting from fill, sq. cut end	0.5
	Headwall or Headwall with Wingwalls	
	Socket end of pipe (groove-end)	0.2
	Square-edge	0.5
	Rounded (radius = 1/12D)	0.2
	Mitered to conform to fill slope	0.7
	End-Section conforming to fill slope*	0.5
	Beveled edges, 33.7° or 45° bevels	0.2
	Side- or slope-tapered inlet	0.2
Pipe or Pipe Arch, Corrugated Metal	Projecting from fill (no headwall)	0.9
	Headwall or headwall and wingwalls square-edge	0.5
	Mitered to conform to fill slope, paved or unpaved slope	0.7
	End-Section conforming to fill slope*	0.5
	Beveled edges, 33.7° or 45° bevels	0.2
	Side- or slope-tapered inlet	0.2
Box, Reinforced Concrete	Headwall parallel to embankment (no wingwalls)	
	Square-edged on 3 edges	0.5
	Rounded on 3 edges to radius of 1/12 barrel dimension, or beveled edges on 3 sides	0.2
	Wingwalls at 30° to 75° to barrel	
	Square-edged at crown	0.4
	Crown edge rounded to radius of 1/12 barrel dimension, or beveled top edge	0.2
	Wingwall at 10° to 25° to barrel	
	Square-edged at crown	0.5
	Wingwalls parallel (extension of sides)	
	Square-edged at crown	0.7
	Side- or slope-tapered inlet	0.2

Table 3-1: Entrance Loss Coefficients

* Note: "End Section conforming to fill slope," made of either metal or concrete, are the sections commonly available from manufacturers. From limited hydraulic tests they are equivalent in operation to a headwall in both <u>inlet</u> and <u>outlet</u> control. Some end sections, incorporating a <u>closed</u> taper in their design have a superior hydraulic performance.

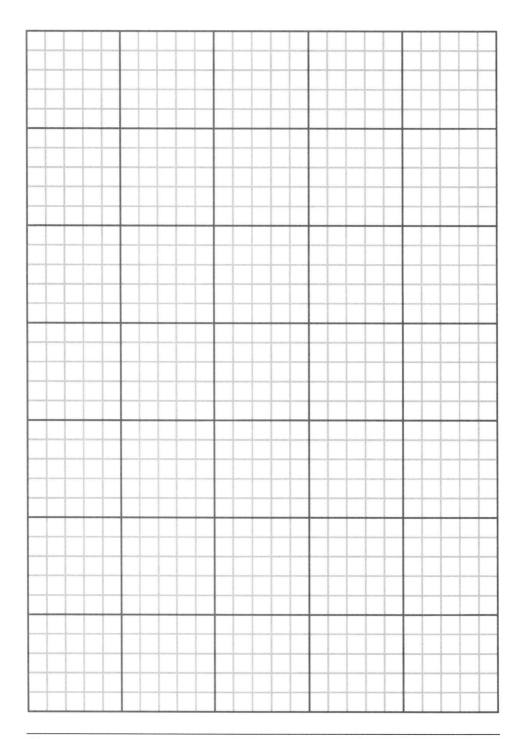

Exit Minor Loss

The exit loss is an expansion loss, which is a function of the change in velocity head that occurs at the discharge end of the culvert. In culvert hydraulics this sudden expansion loss is expressed as:

$$H_o = 1.0 \left[\frac{V^2}{2g} - \frac{V_d^2}{2g} \right]$$

where V_d is the velocity of the outfall channel (ft/s or m/s)
          V is the velocity just inside the end of the culvert barrel (ft/s or m/s)
          g is gravitational acceleration (ft/s^2 or m/s^2)

When the discharge velocity is negligible (as for a pond or slow moving channel) the exit loss is equal to the barrel velocity head.

3.3   Inlet Control Hydraulics

When a culvert is operating under inlet control conditions, the hydraulic control section is the culvert entrance itself. This means that the friction and minor losses within the culvert are not as significant as the losses caused by the entrance constriction.

Since the control section of culverts operating under inlet control conditions is at the upstream end of the culvert barrel, critical depth generally occurs at or near the inlet and flows downstream of the inlet are in the supercritical flow regime. The hydraulic profile and outlet velocities are determined using frontwater gradually varied flow techniques.

Three types of inlet control hydraulics are in effect over a range of culvert discharges:

- *Unsubmerged* - For low discharge conditions, the culvert entrance acts as a weir. The hydraulics of weir flow are governed by empirical working equations developed as a result of model tests.

- *Submerged* - When the culvert entrance is fully submerged, the culvert entrance is assumed to be operating as an orifice.

- *Transitional* - This flow type occurs in the poorly defined region just above the unsubmerged zone, but below the fully submerged zone.

Unsubmerged Flow

There are two equation forms for unsubmerged (weir) flow. The first is based on the specific head at critical depth (with correction factors). The second of these equations is more closely related to a weir equation. Either equation form will produce adequate

---

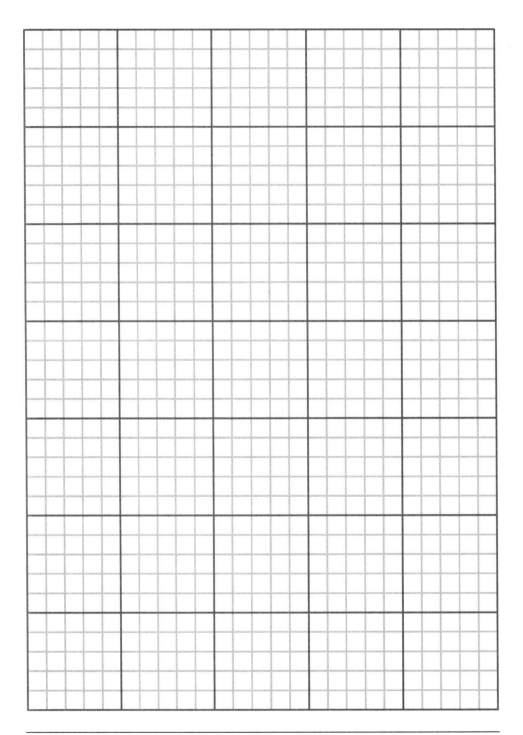

Haestad Methods, Inc.

results, but the second equation is more commonly used during hand calculations because it is easier to apply.

Note that these equations were developed by the United States Federal Highway Administration, and as such are only intended for use in U.S. standard units. Metric calculations should be converted into the correct units before applying these equations.

Unsubmerged, Form 1

$$\frac{HW_i}{D} = \frac{H_c}{D} + K\left[\frac{Q}{AD^{0.5}}\right]^M - 0.5S$$

Unsubmerged, Form 2

$$\frac{HW_i}{D} = K\left[\frac{Q}{AD^{0.5}}\right]^M$$

where HW_i is the headwater depth above the control section invert (ft)
       D   is the interior height of the culvert barrel (ft)
       H_c is the specific head at critical depth, $y_c + V_c^2/2g$ (ft)
       Q   is the culvert discharge (ft³/s)
       A   is the full cross-sectional area of the culvert barrel (ft²)
       S   is the culvert barrel slope
       K and M are constants from the following table

These equations are applicable up to about $Q/AD^{0.5} = 3.5$. When using the first equation with mitered inlets, use a slope correction factor of +0.7S instead of -0.5S.

Submerged Flow

There is only one equation form for submerged (orifice) flow, as follows:

$$\frac{HW_i}{D} = c\left[\frac{Q}{AD^{0.5}}\right]^2 + Y - 0.5S$$

where HW_i is the headwater depth above the control section invert (ft)
       D   is the interior height of the culvert barrel (ft)
       H_c is the specific head at critical depth, $y_c + V_c^2/2g$ (ft)
       Q   is the culvert discharge (ft³/s)
       A   is the full cross-sectional area of the culvert barrel (ft²)
       S   is the culvert barrel slope
       c and Y are constants from the following table

This equation for submerged flow is applicable above about $Q/AD^{0.5} = 4.0$. When using this equation with mitered inlets, use a slope correction factor of +0.7S instead of -0.5S.

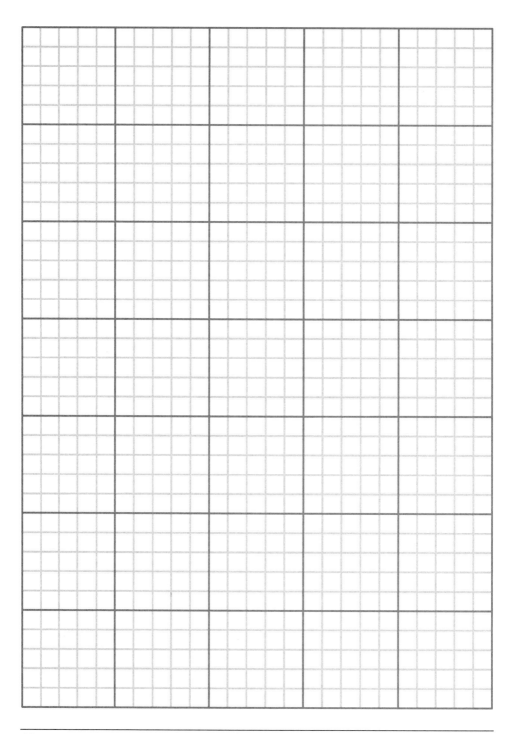

Haestad Methods, Inc.

SHAPE AND	INLET EDGE		UNSUBMERGED		SUBMERGED	
MATERIAL	DESCRIPTION	FORM	K	M	C	Y
Circular	Square edge w/headwall	1	0.0098	2.0	0.0398	0.67
Concrete	Groove end w/headwall		.0078	2.0	.0292	.74
	Groove end projecting		.0045	2.0	.0317	.69
Circular	Headwall	1	.0078	2.0	.0379	.69
CMP	Mitered to slope		.0210	1.33	.0463	.75
	Projecting		.0340	1.50	.0553	.54
Circular	Beveled ring, 45° bevels	1	.0018	2.50	.0300	.74
	Beveled ring, 33.7° bevels		.0018	2.50	.0243	.83
Rectangular	30° to 75° wingwall flares		.026	1.0	.0385	.81
Box	90° and 15° wingwall flares	1	.061	0.75	.0400	.80
	0° wingwall flares		.061	0.75	.0423	.82
Rectangular	45° wingwall flare d=.0430	2	.510	.667	.0309	.80
Box	18° to 33.7° wingwall flare d=.0830		.486	.667	.0249	.83
Rectangular	90° headwall w/¾" chamfers	2	.515	.667	.0375	.79
Box	90° headwall w/45° bevels		.495	.667	.0314	.82
	90° headwall w/33.7° bevels		.486	.667	.0252	.865
Rectangular	¾" chamfers; 45° skewed headwall	2	.522	.667	.0402	.73
Box	¾" chamfers; 30° skewed headwall		.533	.667	.0425	.705
	¾" chamfers; 15° skewed headwall		.545	.667	.04505	.68
	45° bevels; 10°-45° skewed headwall		.498	.667	.0327	.75
Rectangular	45° non-offset wingwall flares	2	.497	.667	.0339	.803
Box	18.4° non-offset wingwall flares		.493	.667	.0361	.806
¾" Chamfers	18.4° non-offset wingwall flares		.495	.667	.0386	.71
	30° skewed barrel					
Rectangular	45° wingwall flares - offset	2	.497	.667	.0302	.835
Box	33.7° wingwall flares - offset		.495	.667	.0252	.881
Top Bevels	18.4° wingwall flares - offset		.493	.667	.0227	.887
C M Boxes	90° headwall	1	.0083	2.0	.0379	.69
	Thick wall projecting		.0145	1.75	.0419	.64
	Thin wall projecting		.0340	1.5	.0496	.57
Horizontal	Square edge w/headwall	1	0.0100	2.0	0.0398	.67
Ellipse	Groove end w/headwall		.0018	2.5	.0292	.74
Concrete	Groove end projecting		.0045	2.0	.0317	.69
Vertical	Square edge w/headwall	1	.0100	2.0	.0398	.67
Ellipse	Groove end w/headwall		.0018	2.5	.0292	.74
Concrete	Groove end projecting		.0095	2.0	.0317	.69
Pipe Arch	90° headwall	1	.0083	2.0	.0379	.69
18" Corner	Mitered to slope		.0300	1.0	.0463	.75
Radius CM	Projecting		.0340	1.5	.0496	.57
Pipe Arch	Projecting	1	.0296	1.5	.0487	.55
18" Corner	No Bevels		.0087	2.0	.0361	.66
Radius CM	33.7° Bevels		.0030	2.0	.0264	.75
Pipe Arch	Projecting	1	.0296	1.5	.0487	.55
31" Corner	No Bevels		.0087	2.0	.0361	.66
Radius CM	33.7° Bevels		.0030	2.0	.0264	.75
Arch CM	90° headwall	1	.0083	2.0	.0379	.69
	Mitered to slope		.0300	1.0	.0463	.75
	Thin wall projecting		.0340	1.5	.0496	.57
Circular	Smooth tapered inlet throat	2	.534	.555	.0196	.89
	Rough tapered inlet throat		.519	.64	.0289	.90
Elliptical	Tapered inlet-beveled edges	2	.536	.622	.0368	.83
Inlet Face	Tapered inlet-square edges		.5035	.719	.0478	.80
	Tapered inlet-thin edge projecting		.547	.80	.0598	.75
Rectangular	Tapered inlet throat	2	.475	.667	.0179	.97
Rectangular	Side tapered-less favorable edges	2	.56	.667	.0466	.85
Concrete	Side tapered-more favorable edges		.56	.667	.0378	.87
Rectangular	Slope tapered-less favorable edges	2	.50	.667	.0466	.65
Concrete	Slope tapered-more favorable edges		.50	.667	.0378	.71

Table 3-2: Coefficients for Inlet Control Design Equations

Computer Applications in Hydraulic Engineering   89

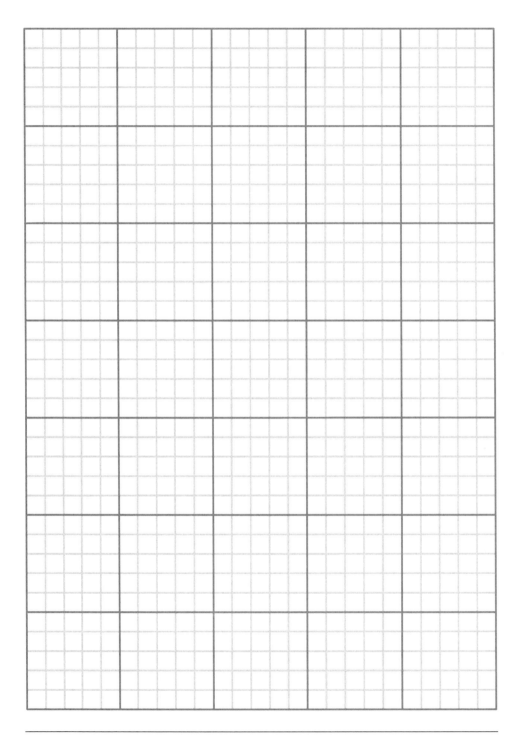

Haestad Methods, Inc.

3.4 Problems

Solve the following problems using the CulvertMaster computer program developed by Haestad Methods (included on the CD accompanying this text).

1. A culvert is 11 meters long and has upstream and downstream inverts of 263.4 and 263.1 meters, respectively. The downstream tailwater elevation is below the downstream pipe invert.

   a) For a K_e of 0.5 and a Manning's "n" value of 0.013, what minimum diameter concrete circular culvert (in mm) is required to pass 1.4 m³/s under a roadway with a maximum allowable headwater elevation of 265.2 meters?

   b) What is the headwater elevation for the selected culvert?

2. An existing 9.73 meter long 560 by 420 mm steel and aluminum arch (n = 0.024) has a headwall with wingwalls. The inverts are 33.11 and 33.09 meters. Assuming no tailwater effects, what is the maximum discharge that can pass through this culvert before the maximum allowable water surface elevation of 34.25 meters is exceeded?

3. Twin 1220 by 910 mm box culverts (n = 0.013, K_e = 0.5) carry 8.5 m³/s along a 31 meter length of pipe constructed at a 1.0 percent slope. The tailwater depth is 0.61 meters.

   a) What is the headwater depth?

   b) Are the culverts flowing under inlet or outlet control conditions?

   c) What would the headwater depth and flow regime be if the flow rate is 380 cfs?

4. A horizontal concrete ellipse pipe (n = 0.013, K_e = 0.5) is required to carry 65 cfs along a 100 foot length at a 3.5 percent slope. Assume there is a free outfall, and a maximum allowable headwater of 4.7 feet.

   a) What minimum size pipe (in inches) is required?

   b) What minimum size vertical concrete ellipse would be required?

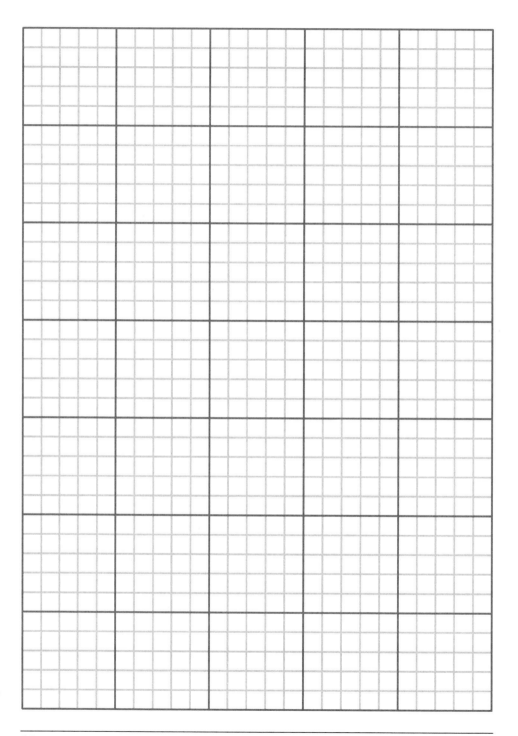

Haestad Methods, Inc.

5. A 12.2 meter long 920 by 570 mm concrete arch pipe (n = 0.013, K_e = 0.2) constructed at a 0.8 percent slope carries 1.84 m^3/s.

   a) If there is a constant tailwater depth of 0.3 meters, what is the headwater depth for both inlet and outlet control conditions?

   b) Is the culvert flowing under inlet or outlet control conditions?

   c) What would be the result if the tailwater was 0.5 meters deeper?

6. Triple 3050 by 1830 mm concrete box culverts (n = 0.013, K_e = 0.5) carry 110 m^3/s. The culverts are constructed on a 1.4% slope, and discharge into a pond with a depth of 0.8 meters above the downstream culvert invert. What is the exit velocity from the culverts?

7. Twin culverts are proposed to discharge 6.5 m^3/s. The culverts will be 36.6 meters long and have inverts of 20.1 and 19.8 meters. The design engineer analyzed the following three culvert systems. Which of the following proposed culverts will result in the highest headwater elevation? The lowest?

   a) 1200 mm circular concrete pipes (n = 0.013 and K_e = 0.5)

   b) 1200 x 910 mm concrete box culverts (n = 0.013 and K_e = 0.5)

   c) 1630 x 1120 mm steel and aluminum var CR arches (n = 0.025 and K_e = 0.5)

8. A 40 foot long elliptical pipe (n = 0.013 and K_e = 0.5) will be constructed to carry 80 cfs with inverts 22.6 and 22.1 feet. The tailwater is constant at elevation 24.0. Which pipe will provide a lower headwater elevation, a 38 by 60 inch horizontal ellipse or a 60 by 38 inch vertical ellipse?

9. A circular concrete culvert has a free outfall. The culvert is 60 feet long on a 2% slope, and is 30 inches in diameter. The culvert entrance will project from the embankment. Create a rating table that correlates the culvert discharge to the depth of the headwater, from 0 feet to 6 feet in 0.5 foot increments.

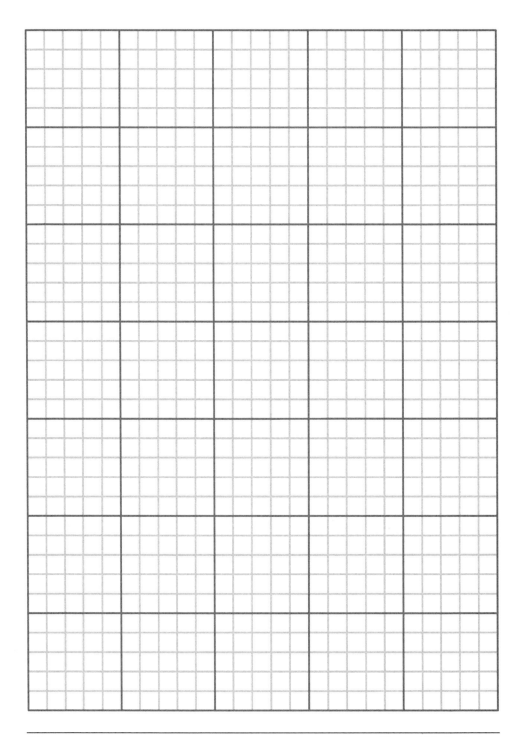

Haestad Methods, Inc.

Chapter 4
Pressure Piping Systems and Water Quality Analysis

4.1 Pressure Systems

Pressure piping network analysis has many applications, including well pumping systems, sewage force-mains, heating and cooling systems, and so forth. The most common application is the one that this chapter primarily deals with: potable water distribution systems.

The basic purpose of a water distribution system is to meet demands for potable water. People use water for drinking, cleaning, gardening, and any number of other uses, and this water needs to be delivered somehow. If designed correctly, the network of interconnected pipes, storage tanks, pumps, and regulating valves provides adequate pressures, adequate supply, and good water quality throughout the system. If incorrectly designed, some areas may have low pressures, poor fire protection, and even health risks.

Water Demands

Just as storm sewer analysis is driven by the watershed runoff, water distribution system analysis is driven by customer demand. Water usage rates and patterns vary greatly from system to system and are highly dependent on climate, culture, and local industry. Every system is different, so the best source for estimating demand comes directly from recorded data for the system.

Metered Demand

Metered demands are often a modeler's best tool, and can be used to calculate average demands, minimum demands, peak demands, and so forth. These data can also be compiled into daily, weekly, monthly, and annual reports that show how the demands are influenced by weather, special events, and other factors.

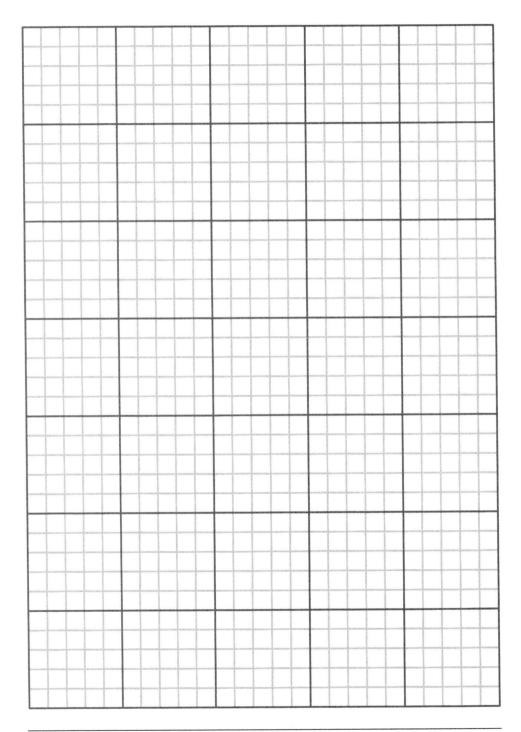

Unfortunately, many systems still do not have complete system metering. For these systems, the modeler is often forced to use other estimation tools (and judgment) to obtain realistic demands.

Demand Patterns

A *pattern* is a function relating water use to time of day. Patterns allow the user to apply automatic time-variable changes within the system. The most common application of patterns is for residential or industrial demands. Diurnal curves are patterns, which relate to the changes in demand over the course of the day, reflecting times when people are using more and less water than average. Most patterns are based on a multiplication factor versus time relationship, whereby a multiplication factor of one represents the base value (which is often the average value). In equation form, this is written as:

$$Q_t = A_t \cdot Q_{Base}$$

where Q_t is the demand at time t
         A_t is a multiplier for time t
         Q_{Base} is the baseline demand

Using a representative diurnal curve for a residence (Figure 4-1), we see that there is a peak in the diurnal curve in the morning as people take showers and prepare breakfast, another slight peak around noon, and a third peak in the evening as people arrive home from work and prepare dinner. Throughout the night, the pattern reflects the relative inactivity of the system, with very low flows compared to the average. (Note that this curve is conceptual and should not be construed as representative of any particular network).

There are two basic forms for representing a pattern; stepwise and continuous. A stepwise pattern is one that assumes a constant level of usage over a period of time and then jumps instantaneously to another level where it remains steady again until the next jump. A continuous pattern is one for which several points in the pattern are known and sections in between are transitional, resulting in a smoother pattern. Notice on the continuous pattern in Figure 4-1 that the value and slope at the start time and end times are the same - a continuity that is recommended for patterns that repeat.

Because of the finite time steps used for calculations, WaterCAD converts continuous patterns into stepwise patterns for use by the algorithms.

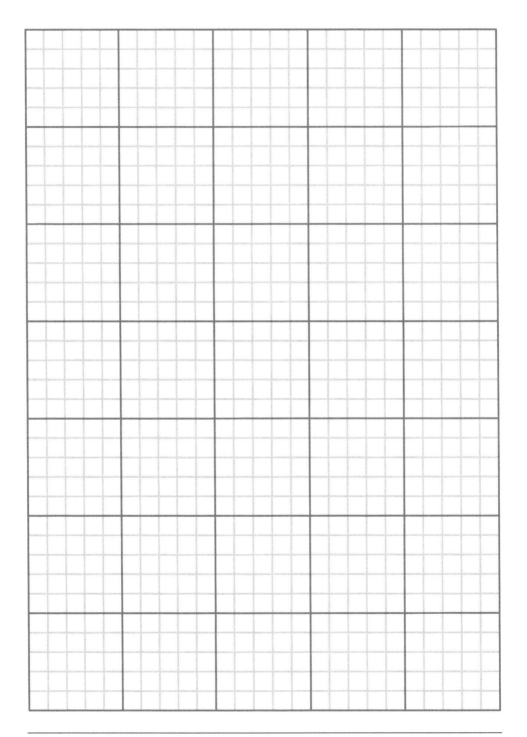

Haestad Methods, Inc.

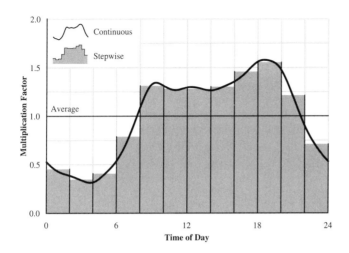

Figure 4-1: Typical Diurnal Curve

4.2 Energy Losses

Friction Losses

The hydraulic theory behind friction losses is the same for pressure piping as it is for open channel hydraulics. The most commonly used methods for determining headlosses in pressure piping systems are the Hazen-Williams equation and the Darcy-Weisbach equation, both discussed in Chapter 1. Many of the general friction loss equations can be simplified and revised, because of the assumptions that can be made for a pressure pipe system:

- Pressure piping is almost always circular, so the flow area, wetted perimeter, and hydraulic radius can all be directly related to diameter.

- Pressure systems flow full (by definition) throughout the length of a given pipe, so the friction slope is constant for a given flowrate. This means that the energy grade and hydraulic grade drop linearly in the direction of flow.

- Since the flowrate and cross-sectional area are constant, the velocity must also be constant. By definition, then, the energy grade line and hydraulic grade line are parallel, separated by the constant velocity head.

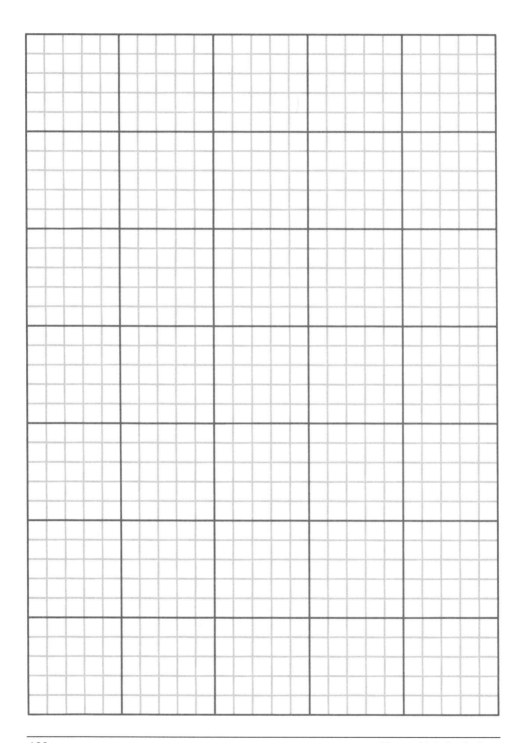

Haestad Methods, Inc.

These simplifications allow for pressure pipe networks to be analyzed much faster than systems of open channels or partial-flow gravity piping. There are several hydraulic components that are unique to pressure piping systems, such as regulating valves, which add some complexity to the analysis.

Minor Losses

Localized areas of increased turbulence cause energy losses within a pipe, creating a drop in the energy and hydraulic grades at that point in the system. These disruptions are often caused by valves, meters, or fittings (such as the pipe entrance in Figure 4-1), and are generally called *minor losses*. These minor losses are often negligible relative to friction losses and are often ignored during analysis.

While the term "minor" is a reasonable generalization for most large-scale water distribution models, these losses may not always be quite as minor as the name implies. In piping systems that contain numerous fittings relative to the total length of pipe, such as heating or cooling systems, the minor losses may actually have a significant impact on the energy losses.

The equation most commonly used for determining the loss in a fitting, valve, meter, or other localized component is:

$$H_m = K \frac{V^2}{2g}$$

where H_m is the minor loss (ft or m)
  K is the minor loss coefficient for the specific fitting
  V is the velocity (ft/s or m/s)
  g is gravitational acceleration (ft/s^2 or m/s^2)

Typical values for the fitting loss coefficient are included in Table 4-1. As can be seen, for similar fittings, the K value is highly dependent on bend radius, contraction ratios, and so forth. Gradual transitions create smoother flow lines and smaller head losses than sharp transitions do, because of the increased turbulence and eddies that form near a sharp change in the flow pattern. This is demonstrated in Figure 4-2, which shows flow lines for a pipe entrance with and without rounding.

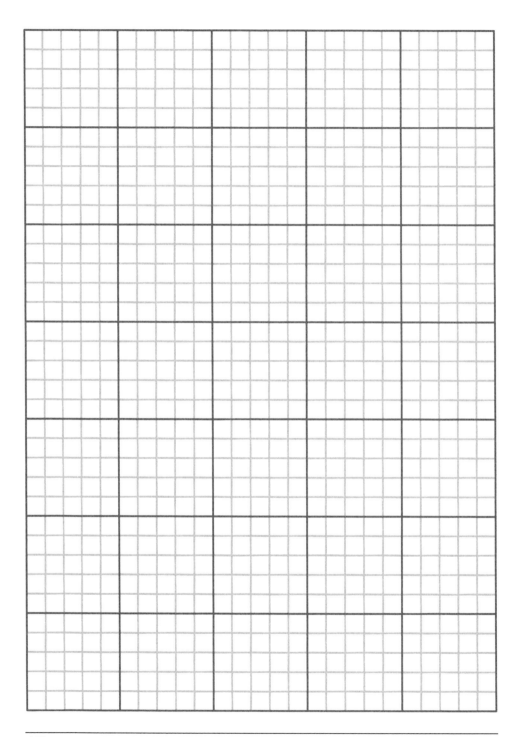

Fitting	K Value
Pipe Entrance	
Bellmouth	0.03 - 0.05
Rounded	0.12 - 0.25
Sharp Edged	0.50
Projecting	0.80
Contraction - Sudden	
$D_2/D_1 = 0.80$	0.18
$D_2/D_1 = 0.50$	0.37
$D_2/D_1 = 0.20$	0.49
Contraction - Conical	
$D_2/D_1 = 0.80$	0.05
$D_2/D_1 = 0.50$	0.07
$D_2/D_1 = 0.20$	0.08
Expansion - Sudden	
$D_2/D_1 = 0.80$	0.16
$D_2/D_1 = 0.50$	0.57
$D_2/D_1 = 0.20$	0.92
Expansion - Conical	
$D_2/D_1 = 0.80$	0.03
$D_2/D_1 = 0.50$	0.08
$D_2/D_1 = 0.20$	0.13

Fitting	K Value
90° Smooth Bend	
Bend radius / D = 4	0.16 - 0.18
Bend radius / D = 2	0.19 - 0.25
Bend radius / D = 1	0.35 - 0.40
Mitered Bend	
$\theta = 15°$	0.05
$\theta = 30°$	0.10
$\theta = 45°$	0.20
$\theta = 60°$	0.35
$\theta = 90°$	0.80
Tee	
Line Flow	0.30 - 0.40
Branch Flow	0.75 - 1.80
Cross	
Line Flow	0.50
Branch Flow	0.75
45° Wye	
Line Flow	0.30
Branch Flow	0.50

Table 4-1: Typical Fitting K Coefficients

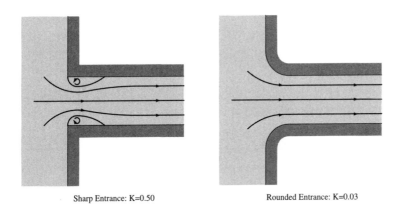

Sharp Entrance: K=0.50   Rounded Entrance: K=0.03

Figure 4-2: Flow Lines in Minor Losses

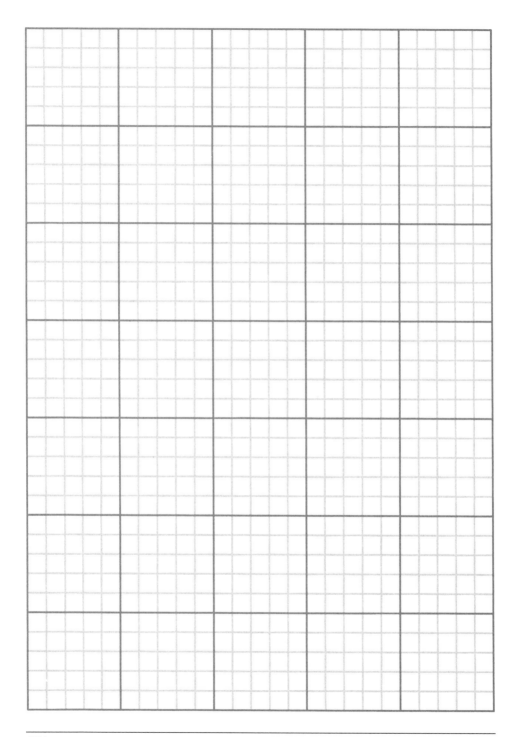

Haestad Methods, Inc.

4.3 Energy Gains - Pumps

Pumps are an integral part of many pressure systems, and are an important part of modeling head change in a network. Pumps add energy (head gains) to the flow, to counteract head losses and hydraulic grade differentials within the system. There are several different types of pump that are used for various purposes; pressurized water systems typically have centrifugal pumps.

A centrifugal pump is defined by its *characteristic curve*, which relates the pump head (head added to the system) to the flow rate. This curve indicates the ability of the pump to add head at different flow rates. To model behavior of the pump system, additional information is needed to ascertain the actual point at which the pump will be operating.

The *system operating point* is based on the point at which the pump curve crosses the *system curve* – the curve representing the static lift and head losses due to friction and minor losses. When these curves are superimposed (as in Figure 4-3) the operating point can be easily located.

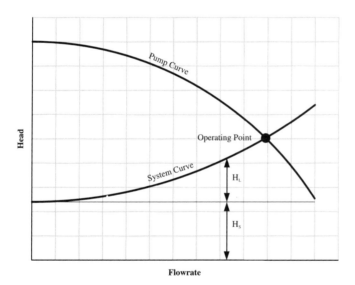

Figure 4-3: System Operating Point

As water surface elevations and demands throughout the system change, the static head (H_s) and head losses (H_L) vary. This changes the location of the system curve, while the pump characteristic curve remains constant. These shifts in the system curve result in a shifting operating point over time.

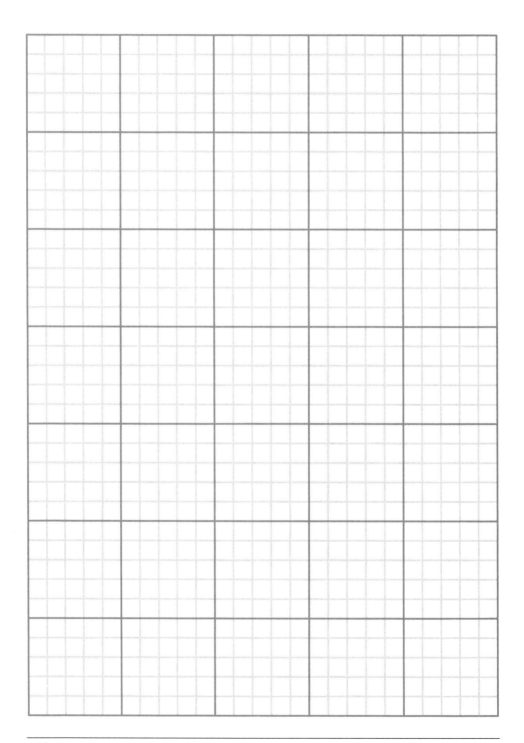

Variable Speed Pumps

A centrifugal pump's characteristic curve is fixed for a given motor speed and impeller diameter, but can be determined for any speed and any diameter by applying the affinity laws. For variable speed pumps, these affinity laws are presented as:

$$\frac{Q_1}{Q_2} = \frac{n_1}{n_2} \quad \text{and} \quad \frac{H_1}{H_2} = \left(\frac{n_1}{n_2}\right)^2$$

where Q is the pump flowrate (ft^3/s or m^3/s)
          H is the pump head (ft or m)
          n is the pump speed (rpm)

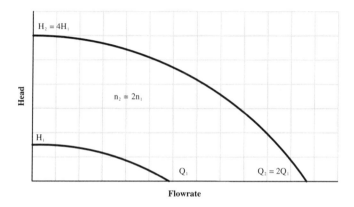

Figure 4-4: Effect of Relative Speed on Pump Curve

Constant Horsepower Pumps

During preliminary studies, the exact characteristics of the pump may not be known. In these cases, the assumption is often made that the pump is adding energy to the water at a constant rate. Horsepower is input as the actual power added to the system and not the rated horsepower of the motor (since there is a loss of efficiency in the motor and motors usually run at less than their rated capacity). Based on power-head-flowrate relationships for pumps, the operating point of the pump can be determined. Although this assumption is useful for some applications, a constant horsepower pump should only be used for preliminary studies.

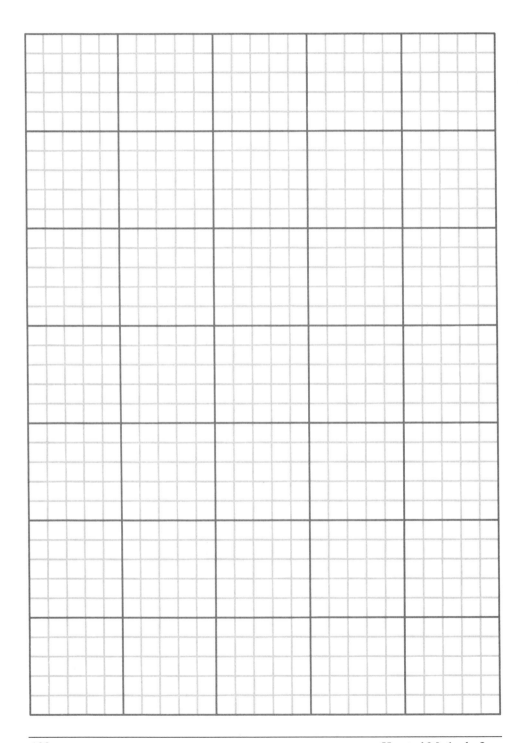

Haestad Methods, Inc.

4.4 Control Valves

There are several types of valves that may be present in a typical pressurized pipe system. These valves have different behavior and different responsibilities, but all valves are used to automatically control parts of the system - opening, closing, or throttling to achieve the desired result.

Check Valves (CV's)

Check valves are used to maintain flow in one direction only, by closing when the flow begins to reverse. When the flow is in the specified direction of the check valve, the valve is considered to be fully open.

Flow Control Valves (FCV's)

A **flow control valve** limits the flowrate through the valve to a specified value, in a specified direction. A flow rate (gpm or l/m) is used to control the operation of a flow control valve. These valves are commonly found in areas where a water district has contracted with another district or a private developer to limit the maximum demand to a value that will not adversely affect the provider's system.

Pressure Reducing Valves (PRV's)

Pressure reducing valves are often used for separate pressure zones in water distribution networks. These valves prevent the pressure downstream from exceeding a specified level, to avoid pressures and flows that could otherwise have undesirable effects on the system. A pressure (psi or kPa) or a hydraulic grade (ft or m) is used to control the operation of a pressure reducing valve.

Pressure Sustaining Valves (PSV's)

Pressure sustaining valves maintain a specified pressure upstream from the valve. Similar to the other regulating valves, these are often used for ensure that pressures in the system (upstream, in this case) will not drop to unacceptable levels. A pressure (psi or kPa) or a hydraulic grade (ft or m) is used to control the operation of a pressure sustaining valve.

Pressure Breaker Valves (PBV's)

Pressure breaker valves create a specified head loss across the valve, and are often used for model components that cannot be easily modeled using standard minor loss elements. A pressure (psi or kPa) or a hydraulic grade (ft or m) is used to control the operation of a pressure breaker valve.

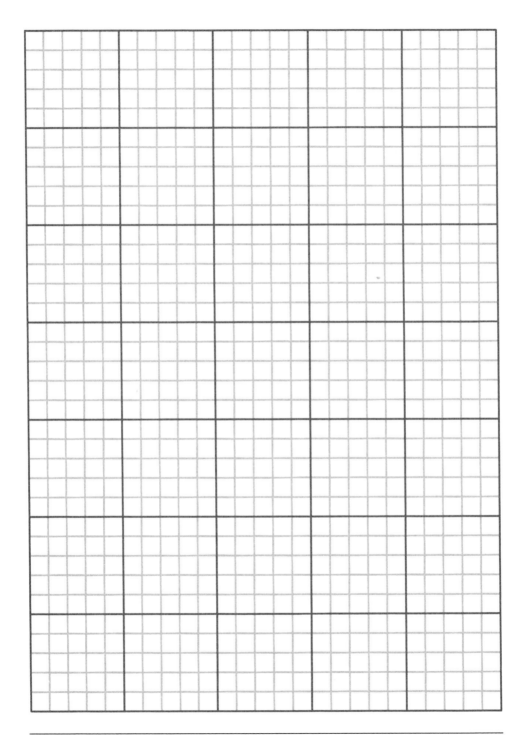

Haestad Methods, Inc.

Throttle Control Valves (TCV's)

Throttle control valves simulate minor loss elements whose head loss characteristics change over time. A headloss coefficient is used to control the operation of a throttle control valve.

4.5 Pipe Networks

In practice, pipe networks consist not only of pipes, but also miscellaneous fittings, services, storage tanks, reservoirs, meters, regulating valves, pumps, and electronic and mechanical controls. For modeling purposes, these system elements are most commonly organized into three fundamental categories:

- ***Junction Nodes***: Junctions are specific points (nodes) in the system located where an event of interest is occurring. This includes points where pipes intersect, where major demands on the system (such as a large industry, a cluster of houses, or a fire hydrant) are located, or critical points in the system where pressures are important for analysis purposes.

- ***Boundary Nodes***: Boundaries are nodes in the system, where the hydraulic grade is known, which define the initial hydraulic grades for any computational cycle. They form the baseline hydraulic constraints used to determine the condition of all other nodes during system operation. Boundary nodes are elements such as tanks, reservoirs and pressure sources. A model must contain at least one boundary node, in order for the hydraulic grade lines and pressures to be accurately calculated.

- ***Links***: Links are system components that connect to junctions or boundaries (such as a pipe) and control the flowrates and energy losses (or gains) between nodes.

An event or condition at one point in the system can affect all other locations in the system. While this complicates the approach that the engineer must take to find a solution, there are some governing principles that drive the behavior of the network, such as the Conservation of Mass and the Conservation of Energy.

Conservation of Mass - Flows and Demands

This principle is a simple one. At any node in the system under incompressible flow conditions, the total volumetric or mass flow in must equal the mass flow out (less the change in storage).

Separating the total volumetric flow into flows from connecting pipes, demands, and storage, we obtain the following equation:

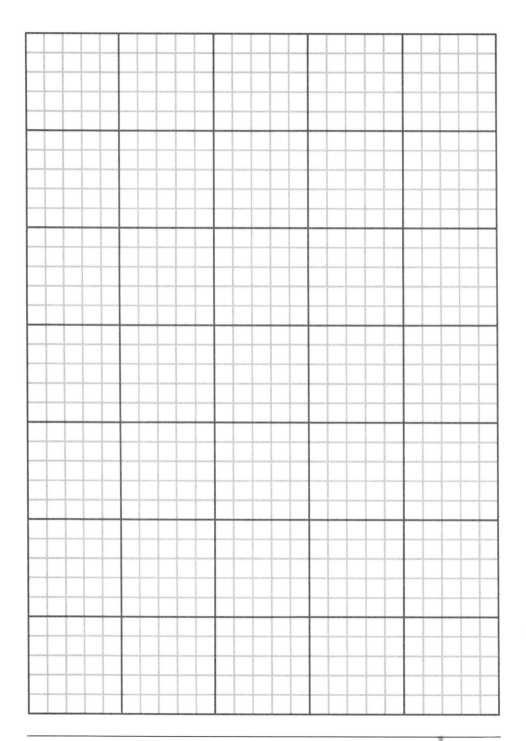

$$\sum Q_{in} \Delta t = \sum Q_{out} \Delta t + \Delta \forall_S$$

where $\sum Q_{in}$ is the total flow into the node
$\sum Q_{out}$ is the total demand at the node
$\Delta \forall_S$ is the change in storage volume
Δt is the change in time

Conservation of Energy

The conservation of energy is also simple conceptually – the head losses through the system must balance at each point. For pressure networks, this means that the total head loss between any two nodes in the system must be the same regardless of what path is taken between the two points. The head loss must be sign consistent with the assumed flow direction (i.e. gain head when proceeding opposite the direction of flow and lose head when proceeding with the flow).

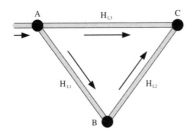

Path from A to C
$$h_{L3} = h_{L1} + h_{L2}$$

Path from A to B
$$h_{L1} = h_{L3} - h_{L2}$$

Loop from A to A
$$0 = h_{L1} + h_{L2} - h_{L3}$$

Figure 4-5 Conservation of Energy

Although the equality can become more complicated with minor losses and controlling valves, the same basic principle can be applied to any path between two points. As shown in Figure 4-5, the combined head loss around a loop must equal zero in order to achieve the same hydraulic grade that was started with.

4.6 Network Analysis

Steady State Network Hydraulics

Steady state analysis is used to determine the operating behavior of a system at a specific point in time, or under steady-state (unchanging) conditions. This type of analysis can be useful to determine the short-term effect of fire flows or average demand conditions on the system.

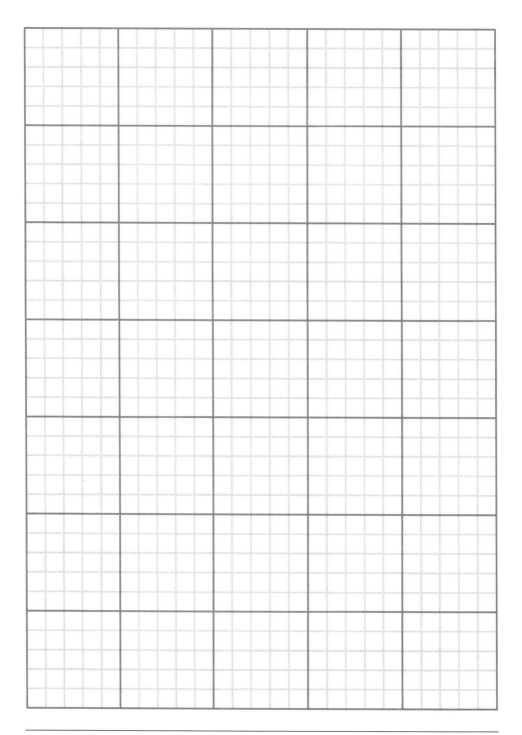

Haestad Methods, Inc.

For this type of analysis, the network equations are determined and solved with tanks being treated as fixed grade boundaries. The results that are obtained from this type of analysis are instantaneous values, and may or may not be representative of the values of the system a few hours, or even a few minutes, later in time.

Extended Period Simulation

An extended period simulation is used to determine the effects on the system over time. This type of analysis allows the user to model tanks filling and draining, regulating valves opening and closing, and pressures and flowrates changing throughout the system in response to varying demand conditions and in response to automatic control strategies formulated by the modeler.

While a steady state model may tell whether or not the system has the capability to meet a certain average demand, an extended period simulation indicates whether or not the system has the ability to provide acceptable levels of service over a period of minutes, hours, or days. Extended period simulations can also be used for energy consumption and cost studies, as well as for water quality modeling.

Data requirements for an extended period simulation are greater than what is needed to model a steady state analysis. In addition to the information required for a steady state model, the user also needs to determine water usage patterns, more detailed tank information, and operational rules for pumps and valves.

4.7 Water Quality Analysis

In the past, water distribution systems were designed and operated with little consideration for water quality, due in part to the difficulty and expense of analyzing a dynamic system. The cost of extensive sampling and the complex interaction between fluids and constituents makes numerical modeling the ideal method for predicting water quality.

To predict water quality parameters, an assumption is made that there is complete mixing across finite distances, such as at a junction node or in a short segment of pipe. Complete mixing is essentially a mass balance, where:

$$C_a = \frac{\Sigma Q_i C_i}{\Sigma Q_i}$$

where C_a is the average (mixed) constituent concentration
          Q_i are the inflow rates
          C_i are the constituent concentrations of the inflows

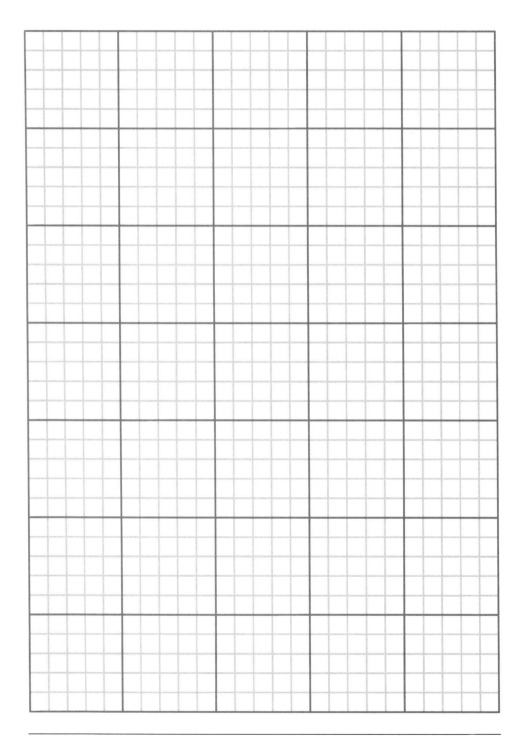

Haestad Methods, Inc.

Age

Water age gives a general indication of the overall water quality at any given point in the system. Age is typically measured from the time that the water enters the system from a tank or reservoir until it reaches a junction.

Water age is computed as:

$$A_j = A_{j-1} + \frac{x}{V}$$

where A_j is the age of water at j-th node
        x is the distance from node j-1 to node j
        V is the velocity from node j-1 to node j

If there are several paths for water to travel to j-th node then the water age is computed as a weighted average using the equation:

$$AA_j = \frac{\sum Q_i \left[AA_i + \left(\frac{x}{V} \right)_i \right]}{\sum Q_i}$$

where AA_j is the average age at node immediately upstream of node j
        Q_i is the flow rate to the j-th node from i-th node

Trace

Identifying the origin of flow at a point in the system is referred to as flow tracking, or trace modeling. In systems that receive water from more than one source, trace studies can be used to determine the percentage of flow from each source at each point in the system. These studies can be very useful in determining the area influenced by an individual source, observing the degree of mixing of water from several sources, and viewing changes in origin over time.

Constituents

Reactions can occur within pipes that cause the concentration of substances to change as the water travels through the system. Based on a conservation of mass for a substance within a link (for extended period simulations only):

$$\frac{\partial c}{\partial t} = V \frac{\partial c}{\partial x} + \theta(c)$$

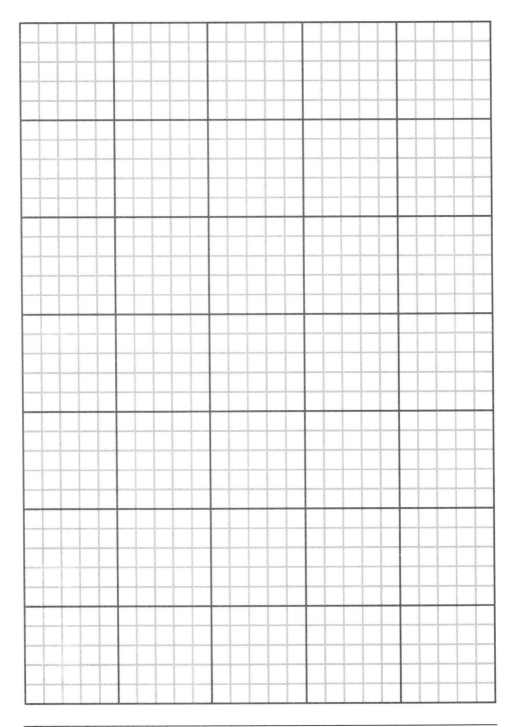

Haestad Methods, Inc.

where   c   is the substance concentration as a function of distance and time
          t   is the time increment
          V   is the velocity
          x   is the distance along the link
          $\theta(c)$ is the substance rate of reaction within the link

In some applications, there is an additional term for dispersion, but this term is usually negligible (indicating plug flow through the system).

Assuming that complete and instantaneous mixing occurs at all junction nodes, additional equations can be expressed for each junction node with the following conservation of mass equation:

$$C_k\big|_{x=0} = \frac{\sum Q_j C_j\big|_{x=L} + Q_e C_e}{\sum Q_j + Q_e}$$

where C_k is the concentration at node k
          j   is all the pipes flowing into node k
          L   is the length of pipe j
          Q_j is the flow in pipe j
          C_j is the concentration in pipe j
          Q_e is the external source flow into node k
          C_e is the external source concentration into node k

Once the hydraulic model has been solved for the network, the velocities and the mixing at nodes are known. Using this information, the water quality behavior can be derived using a numerical method.

Initial Conditions

Just as a hydraulic simulation starts with some amount of water in each storage tank, initial conditions must also be assumed for a water quality analysis of age, trace, or constituent concentration. These initial water quality conditions are usually unknown , and the modeler must estimate these values from field data, a previous water quality model, or some other source of information.

To overcome the problem of unknown initial condition at all locations within the water distribution model, the duration of the analysis must be long enough for the system to reach equilibrium conditions. Note that a constant value does not have to be reached for equilibrium to be achieved; equilibrium conditions are reached when a pattern in age, trace, or constituent concentration is established.

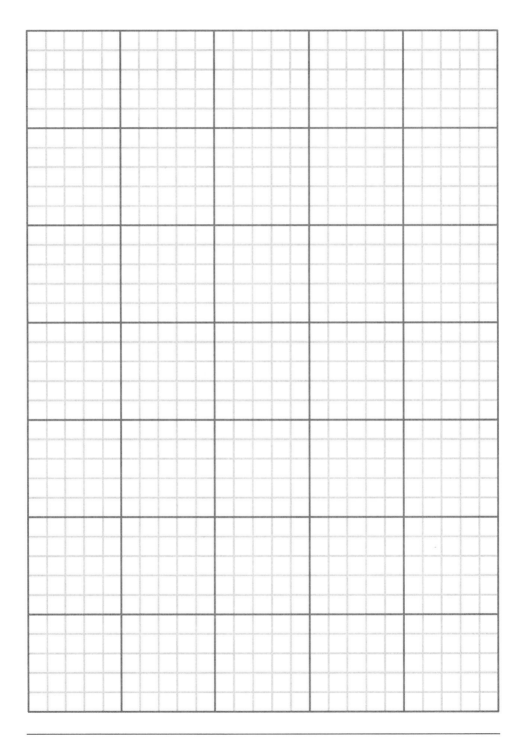

Haestad Methods, Inc.

Equilibrium conditions in pipes are usually reached in a short amount of time, but storage tanks are much slower to reach equilibrium conditions. For this reason, extra care must be taken when setting a tank's initial conditions, to ensure the model's accuracy.

Numerical Methods

There are several theoretical approaches available for solution of water quality models. These methods can generally be grouped as either Eulerian or Lagrangian in nature, depending on the volumetric control approach that is taken. Eulerian models divide the system into fixed pipe segments and then track the changes that occur as water flows through these segments. Lagrangian models also break the system into control volumes, but then track these water volumes as they travel through the system. This chapter presents two alternative approaches for performing water quality constituent analyses.

Discrete Volume Method

The Discrete Volume Method (DVM) is an Eulerian approach that divides each pipe into equal segments with completely mixed volumes. Reactions are calculated within each segment, and the constituents are then transferred to the adjacent downstream segment. At nodes, mass and flow entering from all connecting pipes are combined (assuming total mixing). The resulting concentration is then transported to all adjacent downstream pipe segments. This process is repeated for each water quality time step, until a different hydraulic condition is encountered. When this occurs, the pipes are divided again under the new hydraulic conditions and the process continues.

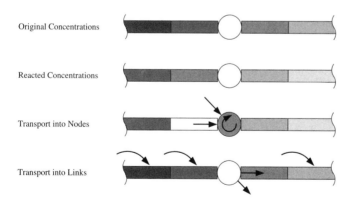

Figure 4-6: Eulerian DVM

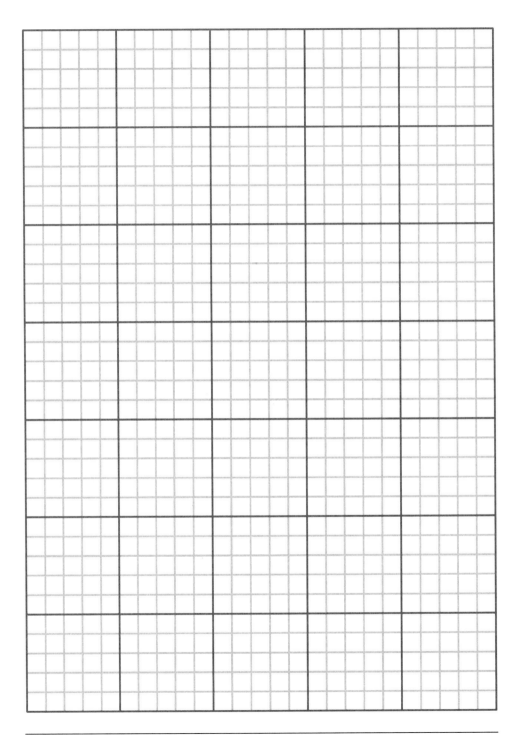

Haestad Methods, Inc.

Time Driven Method

The Time Driven Method (TDM) is an example of a Lagrangian approach. This method also breaks the system into segments, but rather than using fixed control volumes as in Eulerian methods, the concentration and size of the segments are tracked as they travel through the pipes. With each time step, the farthest upstream segment of each pipe elongates as water travels into the pipe, and the furthest downstream segment shortens as water exits the pipe.

Similarly to the Discrete Volume Method, the reactions of a constituent within each segment of the pipe are calculated, and the mass and flow entering each node are summed to determine the resulting concentration. If the resulting nodal concentration is significantly different from the concentration of a downstream pipe segment, a new segment will be created rather than elongating the existing one. These calculations are repeated for each water quality time step, until the next hydraulic change is encountered and the procedure begins again.

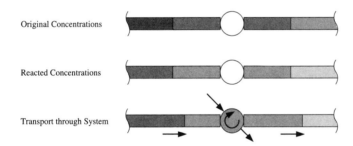

Figure 4-7: Lagrangian TDM

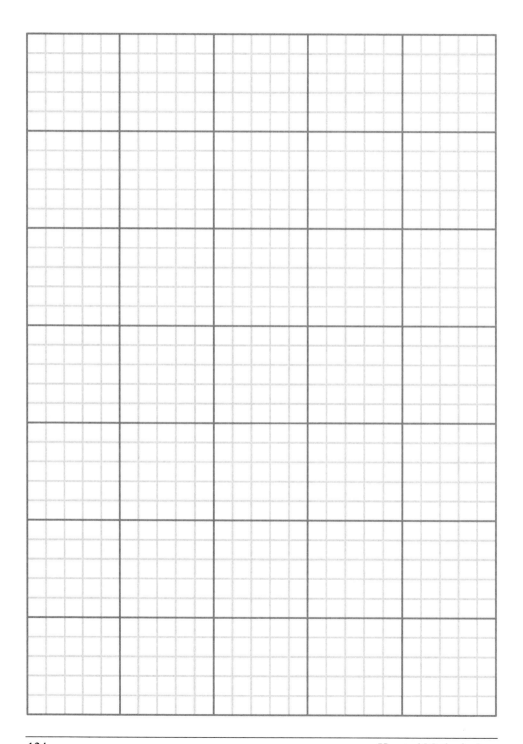

Haestad Methods, Inc.

4.8 Problems

Solve the following problems using the WaterCAD computer program developed by Haestad Methods (included on the CD accompanying this text).

1. The ductile iron pipe network shown below carries water at 20°C. Assume that the junctions are all at elevation = 0 meters, and the reservoir is at 30 meters. Use the Hazen-Williams formula (C = 130) and the following pipe and demand data for a steady state analysis to answer the following questions:

   a) Which pipe carries the lowest discharge, and what is the discharge in liters per minute?

   b) Which pipe has the highest velocity and what is the velocity in m/s?

   c) Calculate the same information using the Darcy-Weisbach equation (k = 0.26 mm).

   d) What effect would raising the reservoir by 20 meters have on the pipe flowrates? What effect would it have on the node hydraulic grades?

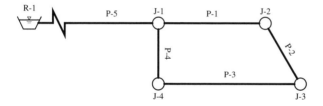

Schematic for Problem 1

Pipe	Diameter (mm)	Length (m)
P-1	150	50
P-2	100	25
P-3	100	60
P-4	100	20
P-5	250	760

Junction	Demand (l/min)
J-1	570
J-2	660
J-3	550
J-4	550

Pipe and Junction Information for Problem 1

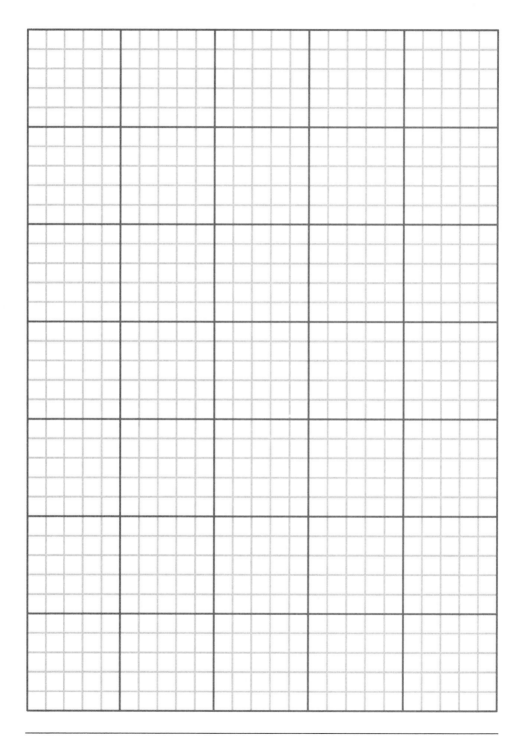

Haestad Methods, Inc.

2. A pressure gage reading of 288 kPa was taken at Junction 5 in the pipe network shown below. Assuming a reservoir elevation of 100 meters, find the appropriate Darcy-Weisbach roughness height (to the hundredth place) that will bring the model into agreement with these field records.

a) What factor gives the best results?

b) What is the calculated pressure at Junction 5 using this factor?

c) Other than the pipe roughnesses, what other factors could cause the model to disagree with field recorded values for flow and pressure?

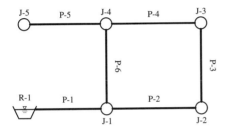

Schematic for Problem 2

Pipe	Diameter (mm)	Length (m)
P-1	250	1525
P-2	150	300
P-3	150	240
P-4	150	275
P-5	150	245
P-6	200	230

Junction	Elevation (m)	Demand (l/min)
J-1	55	950
J-2	49	1060
J-3	58	1440
J-4	46	1175
J-5	44	980

Pipe and Junction Information for Problem 2

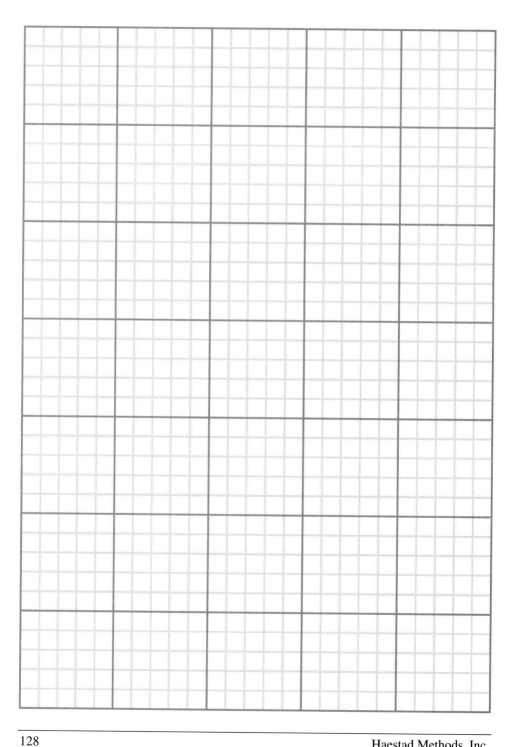

Haestad Methods, Inc.

3. A distribution system is needed to supply water to a resort development for normal usage and also emergency purposes (such as fighting a fire). The proposed system is laid out in the following figure:

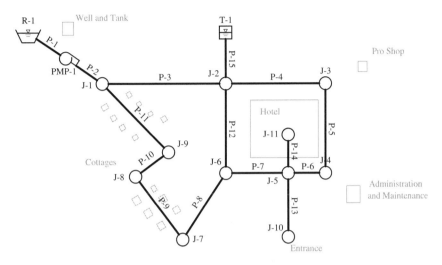

Proposed Network for Problem 3

The source of water for the system is a pumped well. The water is treated and placed in a ground level tank (shown above as a reservoir because of its plentiful supply), which is maintained at a water surface elevation of 210 feet. The water is then pumped from this tank into the rest of the system.

The well system alone cannot efficiently provide the amount of water needed for fire protection, so an elevated storage tank is also needed. The bottom of the tank is at 376 feet (high enough to produce 35 psi at the highest node), and the top is approximately 20 feet higher. To avoid the cost of an elevated tank, this 80-foot diameter tank is located on a hillside, 2000 feet away from the main system. Assume that the tank starts with a water surface elevation of 380 feet.

The pump was originally sized to deliver 300 gpm with enough head to pump against the tank when it's full. Three defining points along the pump curve are as follows: 0 gpm at 200 feet of head; 300 gpm at 180 feet of head; 600 gpm at 150 feet of head. The pump elevation can be assumed to be the same as the elevation at J-1, although the precise pump elevation isn't crucial to the analysis.

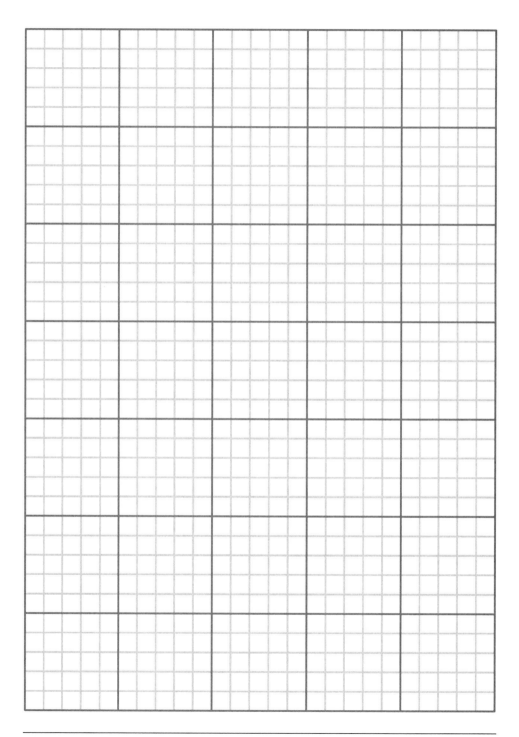

Haestad Methods, Inc.

The system is to be analyzed under several different demand conditions, with minimum and maximum pressure constraints. During normal operations, the junction pressures should all be between 35 psi and 80 psi. Under fire flow conditions, however, the minimum pressure is allowed to drop to 20 psi. Fire protection is being considered both with and without a sprinkler system.

Demand Scenarios: WaterCAD enables you to store several different demand scenarios corresponding to the various conditions (such as average day, peak hour, etc.). This enables you to set up and work with several demand scenarios, without losing any of the input data. For an introduction and more information about scenarios, look in the on-line help system under such keywords as "scenario" or "demand".

Junction	Elevation (ft)	Average Day (gpm)	Peak Hour (gpm)	Minimum Hour (gpm)	Fire with sprinkler (gpm)	Fire without sprinkler (gpm)
J-1	250	0	0	0	0	0
J-2	260	0	0	0	0	0
J-3	262	20	50	2	520	800
J-4	262	20	50	2	520	800
J-5	270	0	0	0	0	800
J-6	280	0	0	0	0	800
J-7	295	40	100	2	40	40
J-8	290	40	100	2	40	40
J-9	285	0	0	0	0	0
J-10	280	0	0	0	360	160
J-11	270	160	400	30	160	160

Junction Information for Problem 3

Pipe Network: The pipe network consists of the pipes listed in the table below. The diameters shown are based on the preliminary design, and may or may not be adequate for the final design. For all pipes, use ductile iron material with a Hazen-Williams C-factor of 130.

Pipe	Diameter (in)	Length (ft)	Pipe	Diameter (in)	Length (ft)
P-1	8	20	P-9	6	400
P-2	8	300	P-10	6	200
P-3	8	600	P-11	6	500
P-4	6	450	P-12	8	500
P-5	6	500	P-13	6	400
P-6	6	300	P-14	6	200
P-7	8	250	P-15	10	2000
P-8	6	400			

Pipe Information for Problem 3

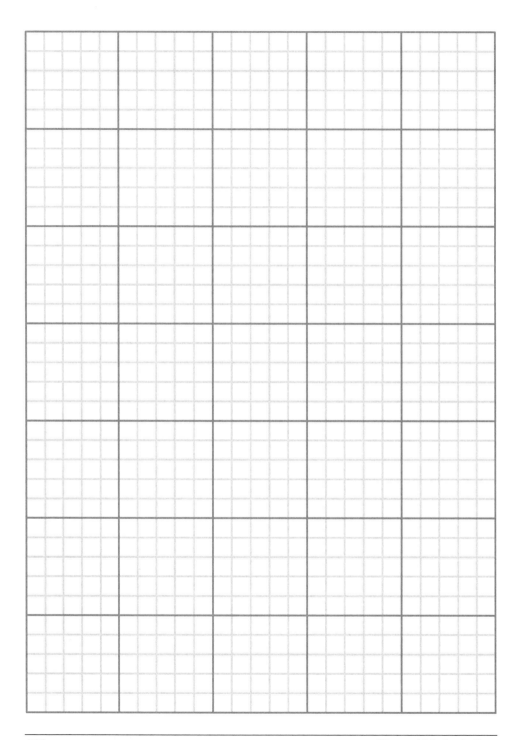

Haestad Methods, Inc.

To help keep track of important system characteristics (like maximum velocity, low pressure, etc.) you may find it helpful to keep a table like the one shown below:

Variable	Average Day	Peak Hour	Minimum Hour	Fire with sprinkler	Fire, without sprinkler
Node w/ low pressure					
Low pressure (psi)					
Node w/ high pressure					
High pressure (psi)					
Pipe w/ max. velocity					
Max. velocity (ft/s)					
Tank in/out flow (gpm)					
Pump discharge (gpm)					

Results Summary for Problem 3

Another way to quickly determine the performance of the system is to color code the pipes according to some indicator. For hydraulic design work, a good indicator is often the velocity of the pipes. Pipes consistently flowing below 0.5 ft/s may be oversized. Pipes with velocity over 5 ft/s are fairly heavily stressed, and those with velocities above 8 ft/s are usually bottlenecks in the system under the flow pattern. Use the following color-coding. When you have defined color coding, place a legend in the drawing.

Max. Velocity (ft/s)	Color
0.5	Magenta
2.5	Blue
5.0	Green
8.0	Yellow
20.0	Red

Color Coding Range for Problem 3

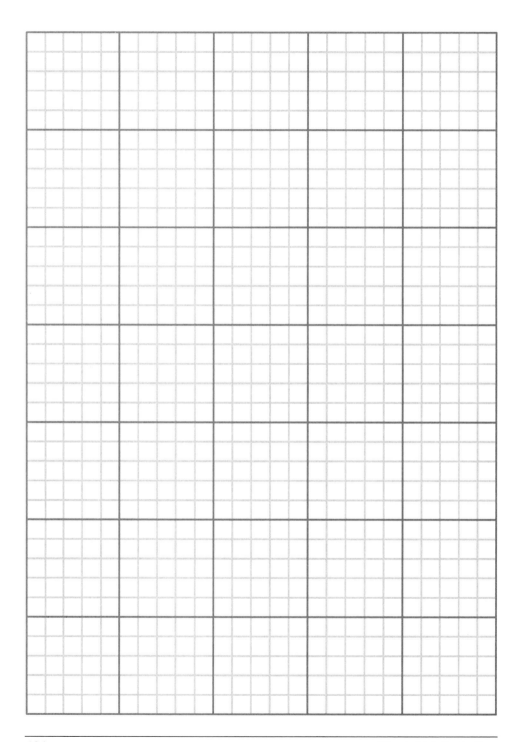

Haestad Methods, Inc.

a) Fill in or reproduce the table on the previous page after each run, to get a feel for some of the key indicators during various scenarios.

b) For the average day run, what was the pump discharge?

c) If the pump has a best efficiency point at 300 gpm, what can you say about its performance on an average day?

d) For the peak hour run, the velocities are fairly low. Does this mean you have oversized the pipes? Why?

e) For the minimum hour run, what was the highest pressure in the system? Why would you expect the highest pressure during the minimum hour demand?

f) Was the system (as currently designed) acceptable for the fire flow case with the sprinkled building? On what did you base this decision?

g) Was the system (as currently designed) acceptable for the fire flow case with all the flow provided by hose streams (no sprinklers)? How would you modify the system so that it would work?

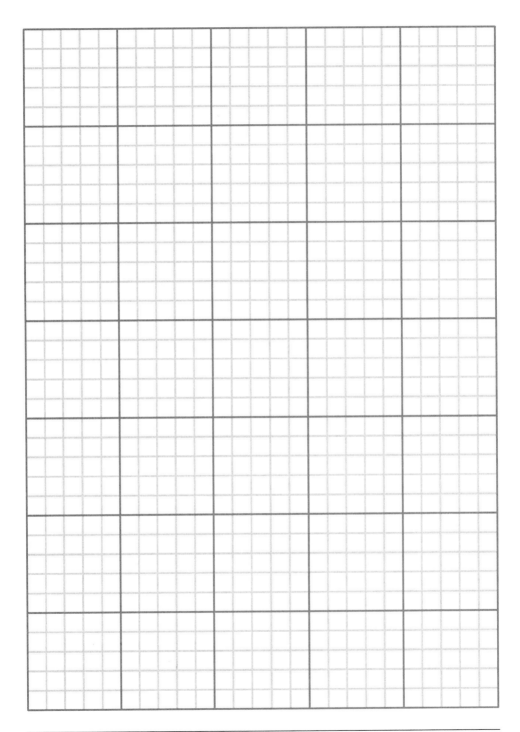

Chapter 5
Using the Software

5.1   General Tips and Common Tools

Because Haestad Methods has adopted object-oriented software techniques, many of the programs share common subsystems, usability, and interface. The following pages summarize some of these features, although you are encouraged to "play" with the software to become more familiar with it. You will find most features to be quite intuitive, and this usability will make you more efficient as a hydraulic modeler.

On-line Help

There has been a definite trend towards paperless documentation, since electronic documentation is easier to keep up-to-date, easily accessible from within the program, and more environmentally friendly. For this reason, you will find that Haestad Methods programs have extensive on-line help systems to guide the user through the analysis and design process. You will probably find that you use Help most often to:

- Get context-sensitive help for an area where you are confused or unsure of nomenclature, data entry, or results

- Search for help topics, such as background theory and equations

- Use "How Do I" to get detailed instructions to complete a task

The following steps assume you have started one of the programs included on the CD and are at the Welcome window, the first window in each of the programs. After you are in a program, you can access Help by clicking on the Help menu.

Context-Sensitive Help

The quickest ways to get help for any task you are doing is to press the F1 key, click on a "Help" button, or right-click with the mouse and select "Help" from the context menu. This gives you *context-sensitive* help, which means that you get help for the window in which you are working.

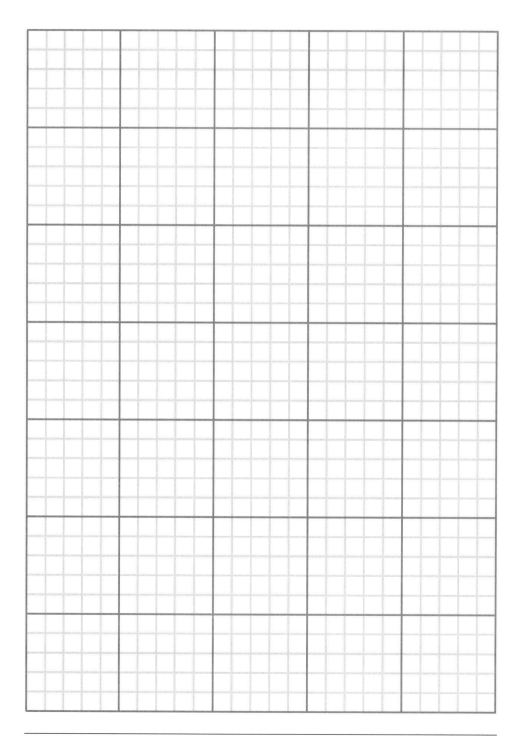

Haestad Methods, Inc.

Search for Help Topics

If the subject you are looking for does not appear immediately when you enter the help system, you may want to search for another topic, based on a keyword. Just click on the "Search" button once the help window appears, and an alphabetical list of items in the index will appear. Type in the topic that you want details for, or browse the list for the subject. If the keyword that you were looking for doesn't appear, don't give up right away - try a few other related words. The index of topics is intended to be complete, but there may be specific terms that are discussed in a slightly different nomenclature.

Use 'How Do I'

The "How Do I" section of the help system has step-by-step instructions for common applications of the model. You may find it beneficial to read through these topics before using the software, or at least to familiarize yourself with the topics. This way, when a related question arises you will remember where to find the answer quickly.

Pop-ups and Jumps

Certain words in each Help topic are underlined and displayed in a different color. These words are known as *pop-ups* and *jumps*. When you select a pop-up (dotted underline), a definition of the word or phrase will appear on the screen. To close the pop-up window, click anywhere away from it. When you select a jump (solid underline), a related Help topic is displayed.

Graphical User Interface (GUI)

Network models, such as StormCAD and WaterCAD, have a graphical user interface (GUI) to aid the engineer in laying out the connectivity of the system. The Haestad Methods GUI is an "intelligent" drafting environment, meaning that it recognizes the characteristics and behavior of different modeling elements, and maintains connectivity when dragging and dropping elements on-screen.

This is a much better system than ASCII-text based models or models where the graphics are not tightly linked to the modeling elements, because it prevents problems from occurring in the first place. Problems that could arise related to incorrect pipe connection, "hanging" pipes (pipes with a node defined at only one end), typographical errors, and other invalid network configurations are avoided. Models that do not have this type of behavior can be easily "broken", and may take hours or even days to locate the source of the problem.

DXF Files

A drawing exchange file (DXF) is the standard format used for translating from one CAD system, such as AutoCAD, to another, such as MicroStation. DXF files can also be imported into StormCAD and WaterCAD for use as a background drawing. For

---

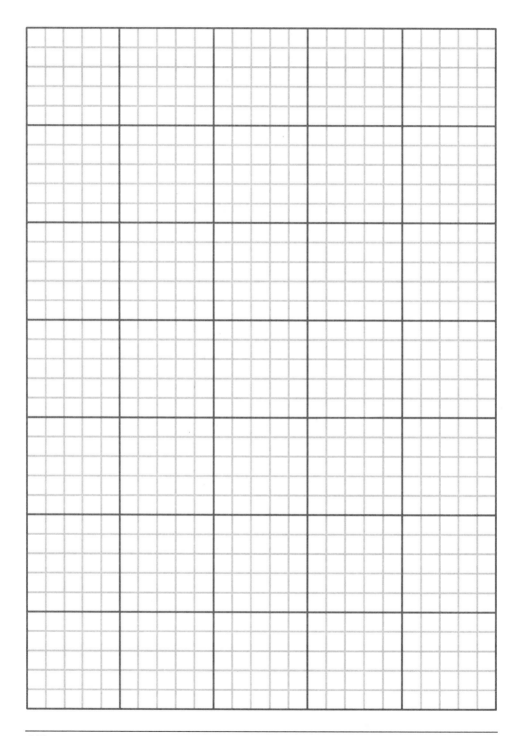

Haestad Methods, Inc.

example, a roadway network plan can be imported to use as a background map for laying out and designing a proposed storm system or water distribution network.

To learn more about importing and using DXF backgrounds, look in the on-line help system of the program (GUI based programs only, such as StormCAD and WaterCAD). You can find topics in the "How Do I" section, or perform a search with a keyword such as 'DXF" or "import".

Scaled or Schematic?

In a scaled plan you set the scale that you want to use as the finished drawing scale for the plan view output. For example, in a drawing with a scale of 1" = 20', one inch on the plan is equal to 20 feet in the real world. Drawing scale is determined based upon engineering judgement and the destination sheet sizes to be used in the final presentations.

A schematic plan is not drawn at any defined scale. There is no connection between dimensions shown on the plan and real world dimensions. There may not even be a correlation between individual dimensions shown on a schematic plan. For example, in a schematic plan the length representing a pipe with a length of 50 meters may be less than, equal to or greater than the line representing a pipe with a length of 20 meters.

In StormCAD and WaterCAD, you may choose either a schematic scale or define the horizontal and vertical distance scale. To get to the drawing scale options choose the Drawing Options item from the Options menu on the menubar.

- Schematic – You enter the pipe length in the Pipe Properties dialog for each pipe. Absolute positioning in the drawing editor is not used.

- Scaled – You enter the horizontal and vertical scale to be used to determine the length of the drawn pipes. The pipe length will be calculated based on the starting and ending coordinates. The length in the drawing editor corresponds to real-world scale. A pipe drawn as 80 feet long in the editor is assigned a length of 80 feet.

Color coding

A way to determine the performance of the system is to color code the pipes according to some indicator. For hydraulic design work a good indicator is velocity in the pipe. For example, pipes consistently flowing below 0.5 ft/s may be oversized. Pipes with velocity over 5 ft/s are fairly heavily stressed while those above 8 ft/s are usually bottlenecks in the system.

Table Manager and Table Customization

From network models, such as StormCAD and WaterCAD, a tabular "datasheet" can be opened and manipulated to aid the modeler in editing the input, and organizing and

---

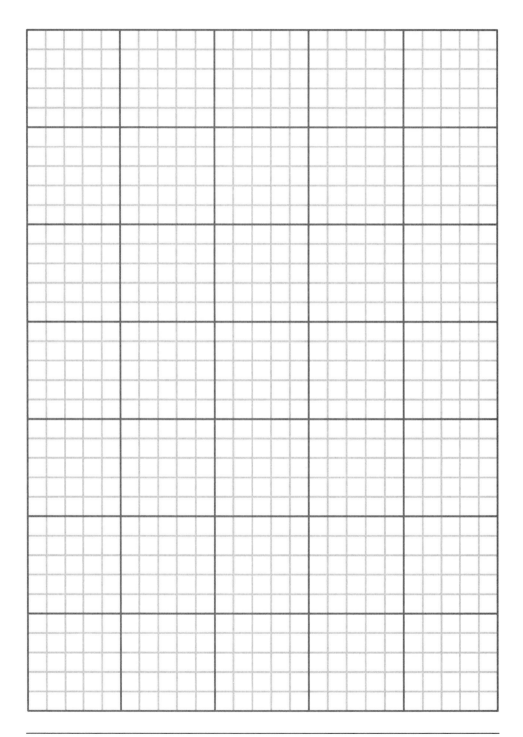

Haestad Methods, Inc.

presenting the results of the model. The tables have several built-in features, which are discussed in more detail in the on-line help system. For more information about customizing and using tables, search in the on-line help for topics such as "table" or "customize". A few of the topics you will find are related to table preferences are briefly given below:

- Changing the Table Title – When you choose to print a Table, the Table name is used as the title for the printed report. You can change the report title by using the table manager to rename the table.

- Adding and Removing Columns – You can add and remove column headings by using the table setup options from the table manager.

- Drag and Drop Column Placement – Select the column that you would like to move by holding down the left mouse button on its column heading, and simply drag the column to its new location and release the mouse button.

- Resizing Columns – In the column heading section of the table, place your mouse over the vertical separator between columns (the cursor shape changes to indicate that you are over the separator). Hold down the left mouse button and drag the mouse to the left or right to set the new column width.

- Changing Column Properties – FlexUnits are available from the tables. To view or edit the properties of any numeric column, right click in the heading area of the column and choose the "... Properties" menu item. The familiar FlexUnits properties will be displayed in the Set Field Options dialog box.

- Changing Column Labels – To change the label of any column, right click on the column heading and choose the Edit Column Label item from the context menu.

- Using Local Units – Local units allow data columns to have fixed units and display precision, regardless of the current unit settings of the model. This can be used to create standardized reports, or it can be used to present the same data side-by-side in different units (especially handy for projects that are transitional between U.S. standard and S.I. units).

In addition to these aesthetic properties of the tables, you can also perform data management operations such as sorting, filtering, and global editing.

Sorting

A table can be sorted in ascending or descending order, according to any variable. This means that the elements can be quickly reorganized based on the alphanumeric labels, pipe flowrates, pipe diameters, hydraulic grade lines, and so forth.

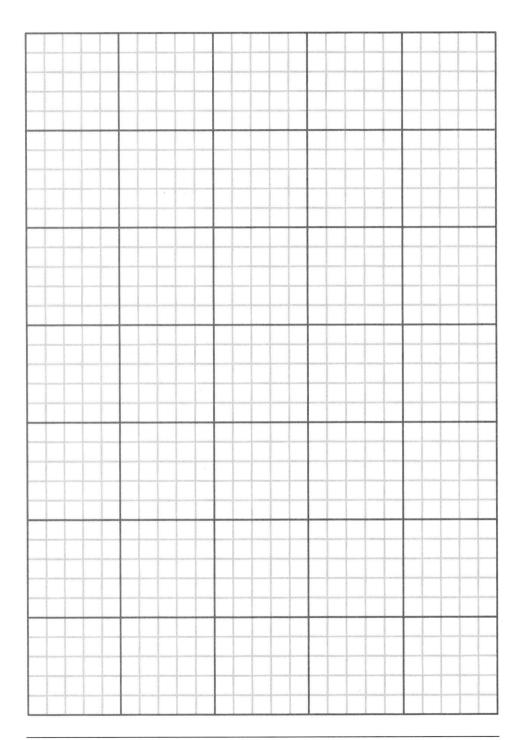

Haestad Methods, Inc.

Filtering

There are often occasions where a modeler may only wish to view a subset of the entire system. For example, there may be concerns in a storm sewer with all pipes that flow more than 10 cubic feet per second, or there may be problems in a water distribution network with any junction nodes that have a pressure below 150 kPa. This process of showing only some of a system's elements is called filtering, and is accomplished by specifying the filtering criterion in the table's Filter dialog box. Each element in the filter list requires three items:

- Column - The table column containing the values to filter on.

- Operator - The operator to use for comparison of the specified filter value against the values in the specified column. Operators include: =, >, >=, <, <=, and <>.

- Value - The comparison value for the filter.

Any number of criteria can be added to a filter. Multiple criteria are implicitly joined with a logical "AND" statement. So, when multiple criteria are defined, only rows that meet all of the specified criteria will be displayed.

The status pane at the bottom of the Tabular Window shows the number of elements displayed and the total number of rows available (e.g. 10 of 20 elements displayed). When a filter is active, the status pane will turn red. Table filtering allows you to perform global editing (see below) on a subset of elements you wish to change.

Global Editing

Factors can be applied to any variable to multiply, divide, add, subtract, or change set all values to meet the modelers needs. Only the elements displayed in the table will be changed, so global editing is often used in conjunction with filtering to obtain the best results.

FlexUnits

From almost anywhere in any of Haestad Methods' Windows programs, the user can set "field options" - the properties of a numerical value. These properties include the units, decimal precision, scientific notation, and the allowable range of values. This feature is referred to as FlexUnits.

To set these options for a field, simply right-click on the field with the mouse, and select "Properties" from the context menu. This will open a dialog where you can set the options that you want:

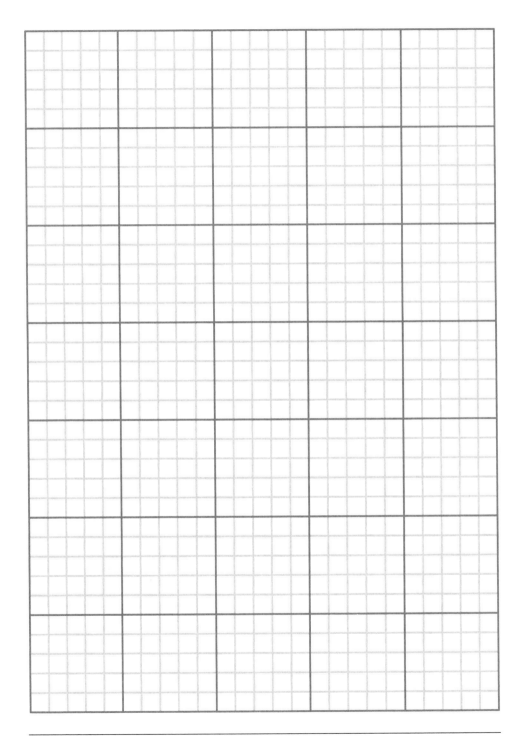

Haestad Methods, Inc.

Units

Because of FlexUnits, Haestad Methods' software offers a wide variety of units for any field where most software only offers a default U.S. unit and a default S.I. unit (at the most). From the Field Options dialog, click on the dropdown menu labeled Units, and simply click on the unit that you want from the list. Using this option allows you to mix and match any units, including both U.S. Standard and S.I. (metric) units.

Once you begin changing units, you will probably notice that some units have multiple representations (for example, psi and lbs/in^2, cfs and ft^3/s). Although these are treated identically by the software, these choices can be made to select the way the units will be presented, both in the program and in your output - another example of just how flexible FlexUnits are.

Display Precision

Display precision can be used to control the number of digits displayed after a decimal point, or the rounding of a numerical value. When the number specified for the display precision is greater than or equal to zero, it specifies the number of digits that are displayed after the decimal point. For example, Pi (3.14159265359) with a decimal precision of 4 would be presented as 3.1416. A negative number for display precision results in a displayed value that is rounded to the nearest power of 10. -1 rounds to 10, -2 rounds to 100, -3 rounds to 1000, and so on. For example, the number 1,234 with display precision of -2 would be displayed as 1,200.

Note that this only affects the way the numbers are displayed, not their actual values. The internal values stored by the software are still carried out to their maximum decimal precision - they are just displayed differently. This is an important concept to keep in mind, especially when checking calculation by hand or working with values that are lower than the usual decimal precision. For example, be aware that a 0.75 inch diameter pipe with a display precision of 0 will result in a displayed value of 1 inch, even though the calculations will be performed based on the true 0.75 inch diameter.

Scientific Notation

Scientific notation displays the number as a real number multiplied by some power of 10. It is displayed as an integer or real value followed by the letter e and an integer (possibly preceded by a sign). For example, 12,345 could be written in scientific notation as 1.2345 e4, and 0.12345 could be written as 1.2345 e-1.

Scientific notation follows the same display precision rules that are outlined above. To turn scientific notation on or off, just click on the field labeled "Scientific notation". When scientific notation is on, a mark will appear within the box.

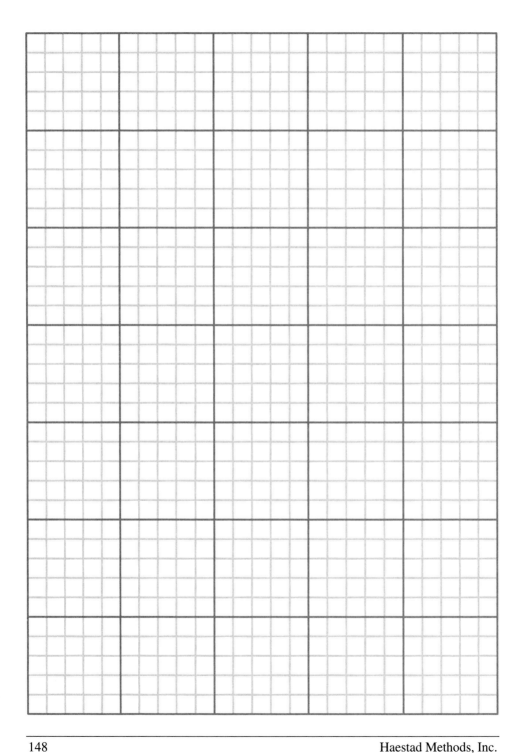

Minimum and Maximum Allowable Values

These options are only available for selected input data, and control the range of values allowed for entry. For example, the user may wish to set minimum or maximum values for roughness coefficients, slope, and so forth.

5.2 FlowMaster

What Does FlowMaster Do?

FlowMaster is an easy-to-use program that helps civil engineers with the hydraulic design and analysis of pipes, ditches, open channels, and more. To do this, FlowMaster computes flows and pressures based on several well-known equations such as Darcy-Weisbach, Manning's, Kutter's, and Hazen-Williams. The program's flexibility allows the user to choose an unknown variable, then automatically compute the solution after entering known parameters. FlowMaster also calculates rating tables, and plots curves and cross sections. You can view the output on the screen, copy it to the Windows clipboard, save it to a file, or print it on any standard printer.

FlowMaster allows you to create an unlimited number of worksheets to analyze uniform sections of pressure pipe or open channel, including irregular sections (such as natural streams or odd-shaped man made sections). FlowMaster does not work with networked systems, such as a storm sewer network or a pressure pipe network. For these types of analysis, see the StormCAD and WaterCAD programs, respectively.

The theory and background used by FlowMaster is presented in more detail in Chapter 1, and also in the FlowMaster on-line help system.

How Can You Use FlowMaster?

FlowMaster replaces solutions such as nomographs, spreadsheets, and "BASIC" programs. Because FlowMaster gives you immediate results, you can quickly generate different output. Not only that, but you perform your hydraulic calculations while taking advantage of Window's many features. Some examples of ways you can use FlowMaster are to:

- Analyze various hydraulic designs

- Evaluate different kinds of flow elements

- Generate professional-looking reports for clients

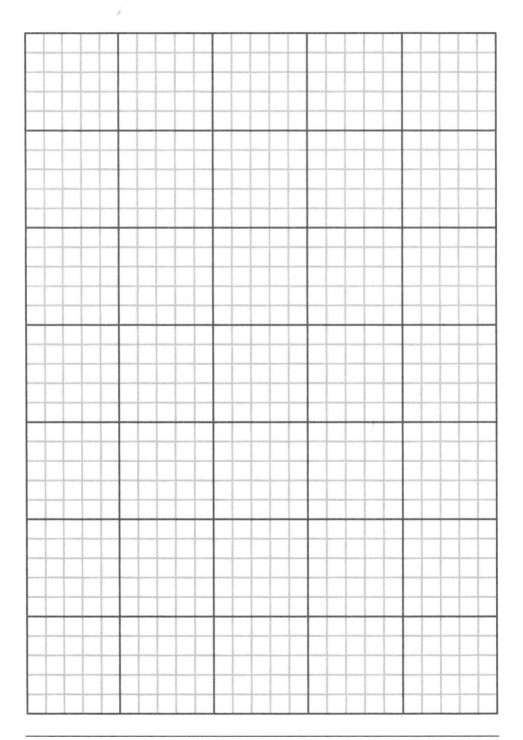

Haestad Methods, Inc.

5.3 StormCAD

What Does StormCAD Do?

StormCAD is an extremely powerful, yet easy-to-use program that helps civil engineers design and analyze storm sewer systems. Just draw your network on the screen using the tool palette, right click on any element to enter data, and click to calculate the network. It's really that easy.

Rainfall information is calculated using rainfall tables, equations, or the National Weather Service's Hydro-35 data. StormCAD also plots Intensity-Duration-Frequency (IDF) curves. You have a choice of conveyance elements including circular pipes, pipe arches, boxes and more. Flow calculations handle pressure and varied flow situations including hydraulic jumps, backwater, and drawdown curves. StormCAD's flexible reporting feature allows you to customize and print the design and analysis results in report format or as a graphical plot.

How Can You Use StormCAD?

StormCAD is so flexible you can use it for all phases of your project, from the feasibility report to the final design drawings. During the feasibility phase, you can use StormCAD to create several different system layouts with an AutoCAD® or MicroStation™ drawing as the background. For the final design, StormCAD lets you complete detailed drawings with notes that can be used to develop construction plans. In summary, you can use StormCAD to:

- Design storm sewer systems

- Analyze various design scenarios for storm sewer systems

- Import and export AutoCAD® and MicroStation™ .DXF files

- Predict rainfall runoff rates

- Generate professional-looking reports for clients

- Generate plan and profile plots of the network

The theory and background used in StormCAD are presented in more detail in Chapter 2, and also in the StormCAD on-line help system.

Analysis and Design

There are three basic methods used to solve a storm sewer system, as shown in the following three topics:

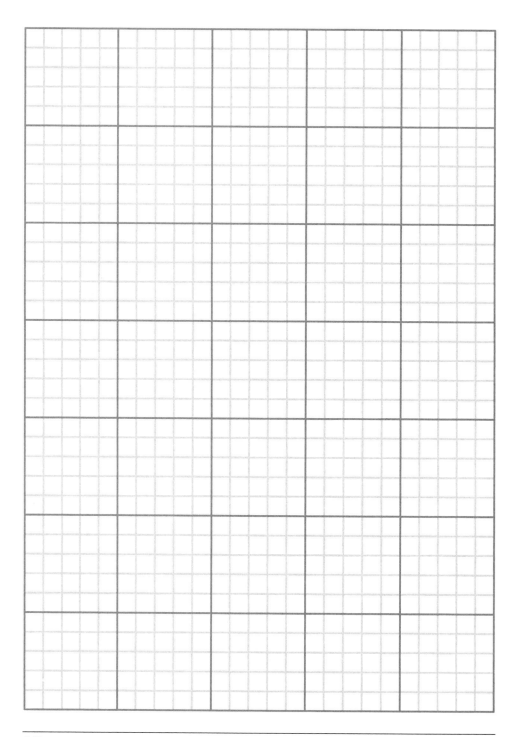

Haestad Methods, Inc.

Fixed Inverts and Diameters

This is used for situations where the engineer already has a set of pipe conditions, and wishes to analyze the system without adjusting these characteristics. Examples of this type of calculation include the analysis of an existing storm sewer, or a design review based on another engineer's proposed piping.

Fixed Inverts, Solve for Diameters

In design situations where utilities or other site constraints may affect the chosen pipe alignment, the engineer may wish to assume pipe inverts and have the model recommend pipe sizes to adequately convey the flow.

Solve for Inverts, Solve for Diameters

A proposed storm system can be designed solving for both invert and pipe diameter. Constraints must be set for pipe diameter, velocity and pipe slope. This is the method that is generally used when designing a new storm sewer system.

Profiles

StormCAD also includes an option to automatically generate storm sewer profiles - longitudinal plots of the storm sewer. Profiles allow the design engineer, the reviewing agency, the contractor and others to visualize the storm system, and are useful for several purposes, including viewing the hydraulic grade line and determining if the proposed storm sewer is in conflict with other existing or proposed underground utilities.

5.4 CulvertMaster

What Does CulvertMaster Do?

CulvertMaster is a program that helps civil engineers design and analyze culvert hydraulics. Just click a button to create a new worksheet; enter data in clearly labeled fields with full context-sensitive help; and click to calculate. You can solve for most hydraulic variables, including culvert size, flow, and headwater. It also allows you to plot and generate output for rating tables showing computed flow characteristics.

You have a choice of culvert barrel shapes including circular pipes, pipe arches, boxes and more. Flow calculations handle pressure and varied flow situations including backwater and drawdown curves. CulvertMaster's flexible reporting feature allows you to print the results in report format or as a graphical plot.

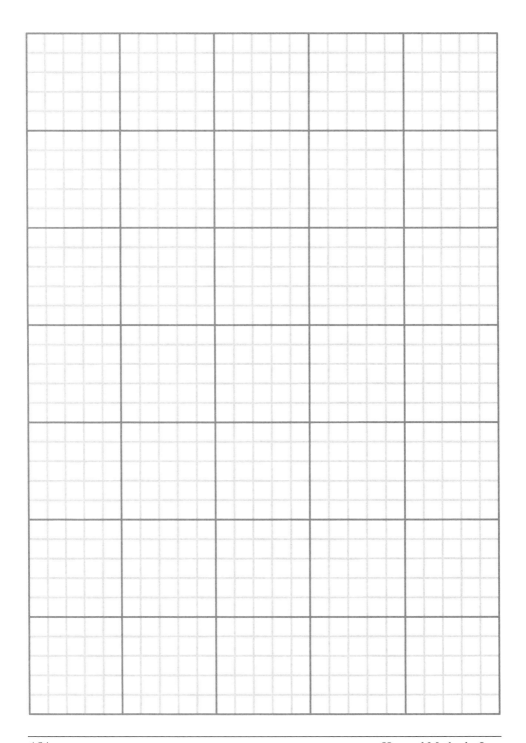

The theory and background used in CulvertMaster is described in detail in Chapter 3, and also in the CulvertMaster on-line help system.

How Can You Use CulvertMaster?

CulvertMaster can design or analyze culverts and compute headwater for a range of flow rates. For a typical CulvertMaster project, you may be interested in several culvert locations and can try several different designs for each location. In summary, you can use CulvertMaster to:

- Size culverts

- Compute and plot rating tables and curves

- View output in both US Customary and SI (metric) units

- Generate professional-looking reports and graphs.

5.5 WaterCAD

What Does WaterCAD Do?

WaterCAD is a powerful, easy-to-use program that helps civil engineers design and analyze water distribution systems. WaterCAD provides intuitive access to the tools you need to model complex hydraulic situations. WaterCAD's sophisticated modeling capabilities can:

- Perform steady state, extended period, and water quality simulations

- Analyze multiple time variable demands at any junction node

- Model pumps using constant horsepower or a multiple point characteristic curve

- Model flow control valves, pressure reducing valves, pressure sustaining valves, pressure breaking valves, throttle control valves

- Model cylindrical and non-cylindrical tanks and constant hydraulic grade source nodes

- Track conservative and non-conservative chemical constituents

- Determine water source and age at any element in the system

The theory and background used in WaterCAD are presented in more detail in Chapter 4, and also in the WaterCAD on-line help system.

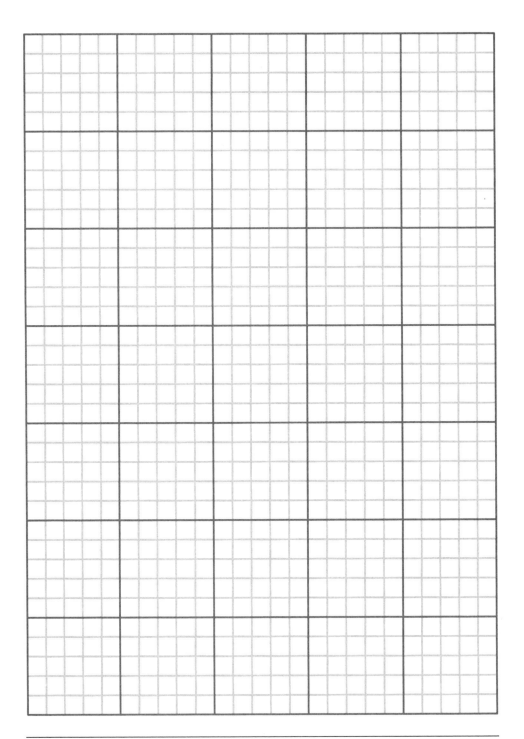

How Can You Use WaterCAD?

WaterCAD can analyze complex distribution systems under a variety of conditions. For a typical WaterCAD project, you may be interested in determining system pressures and flow rates under average loading conditions, peak loading conditions or under fire flow conditions. Extended period analysis tools also allow you to model the system's response to varying supply and demand schedules over a period of time; you can even track chlorine residuals or determine the source of the water at any point in the distribution system.

In summary, you can use WaterCAD for:

- Pipe Sizing

- Pump Sizing

- Master Planning

- Operational Studies

- Rehabilitation Studies

- Water Quality Studies

WaterCAD is a state-of-the-art software tool primarily for use in the modeling and analysis of water distribution systems. While the emphasis is on water distribution systems, the methodology is applicable to any fluid system with the following characteristics:

- Steady or slowly-changing turbulent flow

- Incompressible, Newtonian, single phase fluid

- Full, closed conduits (pressure system)

Examples of systems with these characteristics include potable water systems, sewage force mains, fire protection systems, well pumps, and raw water pumping.

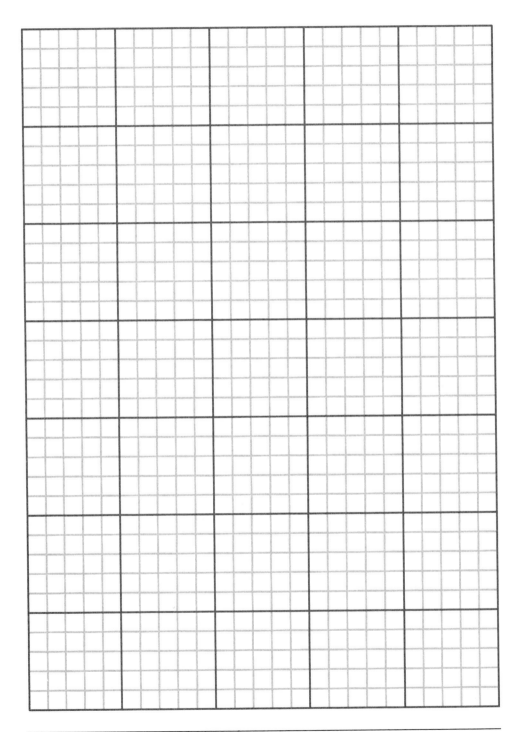

Haestad Methods, Inc.

Bibliography

AASHTO Drainage Manual, American Association of Highway and Transportation Officials, 1991

American Society of Civil Engineers, *Design and Construction of Urban Stormwater Management Systems,* American Society of Civil Engineers, New York, 1982.

American Society of Civil Engineers. *Gravity Sanitary Sewer Design and Construction.* New York: American Society of Civil Engineers; 1982.

Benedict, Robert P., *Fundamentals of Pipe Flow,* John Wiley & Sons, New York, 1980.

Brater, Ernest F. and Horace Williams King, *Handbook of Hydraulics,* McGraw-Hill Book Company, New York, 1976.

Cesario, A. Lee, *Modeling, Analysis, and Design of Water Distribution Systems*, AWWA, 1995

Chow, Ven Te, *Open-Channel Hydraulics,* McGraw-Hill Book Company, New York, 1959.

Clark, R. M., W. M. Grayman, R. M. Males, and A. F. Hess, "Modeling Contaminant propagation in Drinking Water Distribution Systems", *Journal of Environmental Engineering*, ASCE, New York, 1993

Concrete Pipe Design Manual, American Concrete Pipe Association, 1978

D. Earl Jones, Jr., *Design and Construction of Sanitary and Storm Sewers,* ASCE Manual of Practice, No. 37, 1970.

Debo, Thomas N. and Andrew J. Reese, *Municipal Storm Water Management,* CRC Press, Florida, 1995.

Featherstone, R.E. and C. Nalluri, *Civil Engineering Hydraulics,* Granada, New York.

French, Richard H. *Open-Channel Hydraulics.* New York: McGraw-Hill Book Company; 1985.

HEC No. 19, Hydrology, Federal Highway Administration, 1984.

Henderson, F.M., *Open Channel Flow*, MacMillan Publishing Co. Inc., New York, 1966

Hwang, Ned H. C. and Carlos E. Hita, *Hydraulic Engineering Systems,* Prentice-Hall Inc., New Jersey, 1987.

Hydraulics Research Station, "Velocity Equations for Hydraulic Design of Pipes", Metric edition, HMSO, London, 1951 (10/81).

Hydrology, Federal Highway Administration, *HEC No. 19,* 1984.

Males R. M., W. M. Grayman and R. M. Clark, "Modeling Water Quality in Distribution System", *Journal of Water Resources Planning and Management*, ASCE, New York, 1988

Measurement of Peak Discharge at Culverts by Indirect Methods, U. S. Department of the Interior, 1982

Normann, Jerome M., Robert J. Houghtalen, and William J. Johnson, *Hydraulic Design Of Highway Culverts, Hydraulic Design Series No. 5*. U.S. Department Of Transportation, Federal Highway Administration; McLean, Virginia, September 1985

Pilgrim, D.H. *Australian Rainfall and Runoff*. Barton, Australia: The Institute of Engineers; 1987.

Roberson, John A. and Clayton T. Crowe, *Engineering Fluid Mechanics (4ᵗʰ Edition)*, Houghton Mifflin Company, Massachusetts, 1990.

Roberson, John A., John J. Cassidy, and Hanif M. Chaudhry, *Hydraulic Engineering*, Houghton Mifflin Company, Massachusetts, 1988.

Rossman, Lewis A. *et al.*, "Numerical Methods for Modeling Water Quality in Distribution Systems: A Comparison", *Journal of Water Resources Planning and Management*, ASCE, New York, 1996

Rossman, Lewis A., *EPANet User's Manual (AWWA Workshop Edition)*, Risk Reduction Engineering Laboratory, Office of Research and Development, USEPA, Ohio, 1993.

Rossman, Lewis A., R. M. Clark, and W. M. Grayman, "Modeling Chlorine Residuals in Drinking-water Distribution Systems", *Journal of Environmental Engineering*, ASCE, New York, 1994

Sanks, Robert L., *Pumping Station Design*, Butterworth-Heinemann, Inc, Stoneham, Massachusetts, 1989.

Simon, Andrew L. *Practical Hydraulics*. New York: John Wiley & Sons, Inc.; 1976.

Streeter, Victor L., E. Benjamin Wylie, *Fluid Mechanics,* McGraw-Hill Book Company, New York, 1985.

TR-55, *Urban Hydrology for Small Watersheds*, U. S. Department of Agriculture, Soil Conservation Service, Engineering Division, June 1986

Walski, Thomas M., *Water System Modeling Using CYBERNET®*, Haestad Methods, Incorporated, 1993.

Wanielista, Martin P., *Hydrology and Water Quantity Control,* John Wiley & Sons, New York, 1990.

Zipparro, Vincent J. and Hans Hasen, *Davis' Handbook of Applied Hydraulics,* McGraw-Hill Book Company, New York, 1993.

Index

SOFTWARE LICENSE AGREEMENT
IMPORTANT, PLEASE READ CAREFULLY:

This Haestad Methods, Inc. (HMI) End-User License Agreement (Agreement) is a legal agreement between HMI and you (either an individual or a single entity, such as a partnership, corporation, LLC, or other entity) for the HMI software product contained in this package (SOFTWARE). The SOFTWARE includes computer software on associated media and printed materials, and may include on-line or electronic documentation. By installing, copying, or otherwise using the SOFTWARE, you agree to be bound by the terms of this Agreement. If you do not agree with the terms of this Agreement, do not install, copy or use the SOFTWARE, but promptly return the package and unused SOFTWARE to HMI.

GRANT OF LICENSE:

The SOFTWARE is licensed, not sold, from HMI to you. HMI retains ownership of the software and any and all copies that you make of it. This SOFTWARE is licensed for the sole use of the original licensee, at the original location, on a single computer. You may install one copy of the SOFTWARE on a single computer at a single location for use by one person at a time. You may not install the SOFTWARE on a server or networked computer unless you first obtain from HMI a license for each computer or station connected to that server or networked computer that has access to the SOFTWARE. A license for the SOFTWARE may not be shared or used concurrently.

ACADEMIC EDITION LICENSE:

The SOFTWARE is licensed for academic purposes only. Professional or commercial use is strictly prohibited under the terms of this agreement.

SOFTWARE TRANSFER:

This Software License is not transferable. This SOFTWARE is licensed for the sole use of the original licensee, at the original location, on a single computer.

RENTAL:

You may not rent or lease the SOFTWARE.

TERMINATION:

Without prejudice to any other rights, HMI may terminate this Agreement if you fail to comply with its terms and conditions. If HMI notifies you that it has terminated this Agreement, you agree to immediately destroy all copies of the SOFTWARE and all of its component parts.

UPGRADES:

If the SOFTWARE is an upgrade of an older HMI software product, you agree to destroy or return to HMI all copies of the older HMI software product within thirty (30) days of installing the upgrade.

COPYRIGHT:

The SOFTWARE is protected by copyright and other intellectual property law in the United States of America and in other countries by international treaties. Therefore, you may not make or sell copies of the SOFTWARE, except that you may either (a) make one copy of the SOFTWARE solely for backup or archival purposes, or (b) install the SOFTWARE on a single computer and keep the original copy solely for backup or archival purposes. You may not make or sell copies of the printed or on-line materials accompanying the SOFTWARE.

WARRANTY:

For a period of one (1) year after you receive this SOFTWARE (regardless of whether you use the SOFTWARE during the one year period), HMI warrants that the media on which it is contained will not be defective. In the event that during this warranty period the media containing the SOFTWARE is defective, your sole remedy is to contact HMI and HMI in its sole discretion will (a) replace or repair the defective media or (b) refund your money upon your returning to HMI the original and all copies of the SOFTWARE.

DISCLAIMER OF WARRANTIES:

EXCEPT FOR THE EXPRESS WARRANTY STATED ABOVE, THE SOFTWARE IS PROVIDED 'AS IS' AND WITHOUT WARRANTY OF ANY KIND, EXPRESSED OR IMPLIED, INCLUDING BUT NOT LIMITED TO ANY IMPLIED WARRANTIES OF MERCHANTABILITY OR FITNESS FOR A PARTICULAR PURPOSE. THE ENTIRE RISK AS TO THE QUALITY AND PERFORMANCE OF THE SOFTWARE LIES WITH YOU.

LIMITATIONS OF LIABILITY:

HMI SHALL NOT BE LIABLE FOR ANY DAMAGES TO YOU OR ANY OTHER PERSON OR ENTITY IN CONNECTION WITH THE USE OF THIS SOFTWARE. UNDER NO CIRCUMSTANCES WILL HMI BE LIABLE FOR INCIDENTAL, CONSEQUENTIAL OR OTHER DAMAGES, EVEN IF HMI HAS BEEN ADVISED OF THE POSSIBILITY OF SUCH DAMAGES. DO NOT USE THIS SOFTWARE IN ANY WAY OR FOR ANY PURPOSE IF YOU DESIRE HMI TO TAKE ANY LIABILITY FOR ITS USE.

LAW OF THE LAND:

Your rights may vary from state to state, and from country to country. Some of the provisions of this Agreement may not apply.

CONTACT US:

If you have any questions or comments, please contact HMI at the address listed in or on this or the accompanying material.